CLINICAL BIOCHEMISTRY

CLINICAL BIOCHEMISTRY

Homer G. Biggs, Ph.D.

Director, Clinical Chemistry
Department of Clinical Pathology
Indiana University Medical Center
Indianapolis

Geraldine Woodson

B.S., M.T. (ASCP)

Assistant Educational Director
Department of Clinical Laboratory Sciences
School of Community and Allied Health Resources
University of Alabama in Birmingham
Birmingham

Medical Department
HARPER & ROW, PUBLISHERS, INC.
Hagerstown, Maryland
New York, Evanston, San Francisco, London

Clinical Biochemistry. Copyright © 1973 by Harper & Row, Publishers, Inc. All rights reserved. No part of this book may be used or reproduced in any manner whatsoever without written permission except in the case of brief quotations embodied in critical articles and reviews. Printed in The United States of America. For information address Medical Department, Harper & Row, Publishers, Inc., 2350 Virginia Avenue, Hagerstown, Maryland 21740.

First edition

Standard Book Number: 06–140466–7

Library of Congress Catalog Card Number: 72–6319

CONTENTS

Preface   ix

Acknowledgments   xi

CHAPTER 1. CARBOHYDRATES   1
General Description and Nomenclature. Isomerism. Reactions. Hemiacetal Structures. Carbohydrate Derivatives. Disaccharides. Polysaccharides. Importance of Carbohydrates

CHAPTER 2. LIPIDS   24
Neutral Lipids. Phospholipids. Compound Lipids. Sterols

CHAPTER 3. AMINO ACIDS AND PROTEINS   35
Amino Acids. Peptides and Proteins. Electrophoresis. Protein Structures. Denaturation

CHAPTER 4. ENZYMES   56
History. Enzyme Action. Common Enzyme Properties. Activity Measurements. Nomenclature. Classification. Reversibility of Enzyme Reactions. Enzyme Inhibition

CHAPTER 5. NUCLEIC ACIDS   76
Pyrimidine and Purine Bases. Nucleosides. Nucleotides. Polynucleotide Structures. Nucleotide Derivatives

CHAPTER 6. DIGESTION 87
Mouth. Stomach. Small Intestine. Large Intestine. Absorption

CHAPTER 7. CARBOHYDRATE METABOLISM 98
Glycolysis. Other Pathways. Metabolic Controls. Hexose Monophosphate Shunt. Enzymatic Detection of Glucose. Glycogen Storage and Blood Glucose. Enzyme Deficiencies

CHAPTER 8. LIPID METABOLISM 115
Fatty Acid Metabolism. Fatty Acid Synthesis. Glyceride Synthesis. Sterol Synthesis. Ketogenesis. Structural Lipids. Fat Storage. Blood Lipids

CHAPTER 9. AMINO ACID METABOLISM 125
Common Metabolic Reactions. Urea Synthesis. Creatine Metabolism. Metabolism of Phenylalanine and Tyrosine. Tryptophan Metabolism. Summary of Amino Acid Metabolism

CHAPTER 10. NUCLEIC ACID METABOLISM 137
Cellular Division. Role of Nucleic Acids. Purine and Pyrimidine Metabolism

CHAPTER 11. COMMON METABOLIC PATHWAYS 146
Acetyl Coenzyme A. Tricarboxycylic Acid Cycle. Electron Transport System

CHAPTER 12. CLINICALLY IMPORTANT ENZYMES 151
Amylase. Lipase. Serum Glutamate Oxalacetate Transaminase. Creatine Phosphokinase. Lactate Dehydrogenase. Clinical Response of SGOT, CPK, and LD. Phosphatases. Aldolase

CHAPTER 13. EQUILIBRIUM, ACIDS, BASES, AND BUFFERS 165
Equilibrium. Acids and Bases. Buffers

Contents   vii

CHAPTER 14.   ACID–BASE BALANCE   180
Sources of Acids.   Sources of Bases.   Isohydric Mechanism.   Buffer Systems.   pH Regulation: Role of Lungs.   pH Regulation: Role of Kidney.   Bicarbonate and Carbonic Acid Measurements.   pH Measurements.   Partial Pressure.   Oxygen Transport.   Carbon Dioxide Transport.   Methods for Evaluating Acid–Base Balance

CHAPTER 15.   ELECTROLYTES   192
Methods for Measuring Electrolytes.   Electrolyte Interrelationships

CHAPTER 16.   HORMONES   200
Anterior pituitary.   Posterior pituitary hormones.   Thyroid.   Adrenal Cortex.   Adrenal Medulla.   Sex Glands.   Parathyroid Glands.   Serotonin

CHAPTER 17.   VITAMINS   213
Fat-Soluble Vitamins.   Water-Soluble Vitamins.   Vitamin Supplements

CHAPTER 18.   PHOTOMETRY   222
Radiant Energy Spectrum.   Absorbance Photometry.   Absorbance Photometers.   Zero Absorbance and Types of Photometers.   Absorbance Photometry and Quantitation.   Calibration or Standardization Curves.   Molar and Specific Absorptivity.   Colorimeters and Spectrophotometers.   Other Types of Photometry

CHAPTER 19.   QUALITY CONTROL   256
Laboratory Data and Statistics.   Quality Control Techniques

INDEX   275

PREFACE

This brief introduction to clinical chemistry is an outgrowth of teaching manuals prepared for laboratory trainees at the University of Alabama in Birmingham. The intent of this material is to provide the student with basic, general information and to demonstrate that basic chemistry and biochemistry are the foundations of clinical chemistry.

The material has been greatly abridged, but when supplemented with appropriate lectures and laboratory exercises, it can promote a comprehensive understanding of clinical chemistry. Hopefully it will furnish enough information to stimulate continued study.

The principles of some methodology and instrumentation are included, but specifics have been avoided for these constantly changing areas. The continual introduction of new procedures and the rapid development of instrumentation would quickly render obsolete detailed descriptions in these areas. This new book will remain applicable and informative for references and study for years to come.

H.G.B.
G.W.

ACKNOWLEDGMENTS

We wish to express our appreciation to Jon V. Straumfjord, Jr., M.D. for his support and interest. His efforts and enthusiasm for the education of medical laboratory personnel were the impetus for our initial venture in the area of manual development.

We gratefully acknowledge the help given by the staff members of the Section of Clinical Chemistry, Department of Clinical Pathology, University of Alabama Medical Center; also the members of Illustration and Design, Office of Learning Resources for preparation of the illustrations.

The authors give special thanks to Edward A. Sasse, Ph.D. for the time he spent reviewing the manuscript. His constructive ideas and undaunted good humor contributed greatly.

Mrs. Ricki W. Self and Mrs. Kathey W. Giles gave much of their time typing and proofreading the various drafts. We thank them for their patience and understanding.

CHAPTER 1

CARBOHYDRATES

Carbohydrates are composed largely of carbon, hydrogen, and oxygen and are probably the most abundant naturally occurring organic compounds. They are found as constituents of both plants and animals and usually represent stored energy.

GENERAL DESCRIPTION AND NOMENCLATURE

As the name implies, carbohydrates can be represented as hydrates of carbon.

1 $$C_x(H_2O)_y$$

This is not a very meaningful definition because many organic compounds, which are not generally considered to be carbohydrates, can also be written as hydrates of carbon. For example:

2 $$CH_3COOH = C_2(H_2O)_2$$
acetic acid

3 $$HCHO = C(H_2O)$$
formaldehyde

A poly-ol with a carbonyl group is a more satisfactory description of a carbohydrate. Glyceraldehyde represents one of the simplest compounds of this description.

4
$$\begin{array}{c} H \\ | \\ C=O \\ | \\ H-C-OH \\ | \\ H-C-OH \\ | \\ H \end{array}$$
glyceraldehyde

The poly-ol requirement is fulfilled by two hydroxy groups and the carbonyl by an aldehyde functional group. Another simple carbohydrate is dihydroxyacetone. Two hydroxy groups plus a carbonyl functional group in the form of a ketone fulfill the requirements.

5
$$\begin{array}{c} H \\ | \\ H-C-OH \\ | \\ C=O \\ | \\ H-C-OH \\ | \\ H \end{array}$$
dihydroxyacetone

It differs structurally from glyceraldehyde in that the carbonyl functional group is in the form of a ketone. It is evident from examination of these formulas that carbohydrates can exist in two distinct forms: those with an aldehyde group and those with a ketone group. Those bearing an aldehyde group are referred to as aldoses and those with a ketone group as ketoses. The suffix ose designates the compounds as carbohydrates.

An examination of more complex carbohydrates reveals that these compounds can have more than three carbons.

6
$$\begin{array}{c} H \\ | \\ C=O \\ | \\ H-C-OH \\ | \\ H-C-OH \\ | \\ H-C-OH \\ | \\ H \end{array}$$
erythrose

Another type of nomenclature takes this into consideration. It is based on the number of carbons which occur in a continuous chain in the molecule. For instance, glyceraldehyde and dihydroxyacetone are called trioses. This does not allow for a distinction between the aldoses and ketoses. However, by combining the two systems of nomenclature, glyceraldehyde can also be called aldotriose because it has three carbons and an aldehyde group, and the suffix ose indicates it is a carbohydrate. Similarly, dihydroxyacetone can be called a ketotriose: keto indicates that the carbonyl group is a ketone, the tri portion refers to three carbons, the ose ending signifies a carbohydrate. Applying the approach to formula **6** the compound can be called an aldotetrose.

ISOMERISM

Keeping in mind such concepts as sterioisomerism and optical isomers, you will recall that certain carbon compounds have structures which can rotate plane polarized light. These structures contain a carbon bonded to four different groups and have no plane of symmetry.

7

$$\begin{array}{c} A \\ | \\ E-C-B \\ | \\ D \end{array}$$

Theoretically, there are two isomers for each asymmetric carbon, as shown in **8**.

8

$$\begin{array}{c c} A & A \\ | & | \\ E-C-B & B-C-E \\ | & | \\ D & D \end{array}$$

The isomers are mirror images, and they are not superimposable. It is difficult to visualize this concept in a two dimensional illustration. However, with three dimensional models it can be readily demonstrated.

Consider the structure of glyceraldehyde. There is one asymmetric carbon atom, therefore the compound has two optical isomers.

9

$$\begin{array}{c c} O & O \\ \| & \| \\ H-C & H-C \\ | & | \\ H-C-OH & HO-C-H \\ | & | \\ H-C-OH & H-C-OH \\ | & | \\ H & H \end{array}$$

D-glyceraldehyde   L-glyceraldehyde

The prefixes D or L in small capital letters designate configuration and not the direction of rotation. The signs (+) and (−) are used to denote dextrorotation and levorotation, respectively. For glyceraldehyde, the D isomer is dextrorotatory, thus, the complete name would be D(+)-glyceraldehyde. Since the L isomer, for glyceraldehyde, is levorotatory, the correct name would be L(−)-glyceraldehyde.

There is no consistent relationship between structure and rotation as the compound glyceraldehyde might suggest. For example, the isomers of lactic acid are D(−)-lactic acid and L(+)-lactic acid.

10
$$\begin{array}{cc}
\text{O} & \text{O} \\
\parallel & \parallel \\
\text{C—OH} & \text{C—OH} \\
| & | \\
\text{H—C—OH} & \text{HO—C—H} \\
| & | \\
\text{CH}_3 & \text{CH}_3 \\
\text{D(—)-lactic acid} & \text{L(+)-lactic acid}
\end{array}$$

Inspection of the structure of aldotetrose reveals two asymmetric carbon atoms, numbers 2 and 3.

11
$$\begin{array}{c}
\text{O} \\
\overset{1}{\text{C}}\text{—H} \\
| \\
\text{H}\overset{2}{-}\text{C—OH} \\
| \\
\text{H}\overset{3}{-}\text{C—OH} \\
| \\
\text{H}\overset{4}{-}\text{C—OH} \\
| \\
\text{H}
\end{array}$$

Since both of the asymmetric carbon atoms can assume two different configurations, there are four possibile optical isomers.

Problems arise concerning nomenclature when a compound contains more than one asymmetric center. For carbohydrates, the convention specifies that the configuration of the asymmetric carbon farthest from the carbonyl carbon designates the L or D series.

For the aldotetroses, the configuration of carbon 3 is the determinant.

12
$$\begin{array}{cc}
\text{H} & \text{H} \\
| & | \\
\overset{1}{\text{C}}\text{=O} & \text{C=O} \\
| & | \\
\text{HO}\overset{2}{-}\text{C—H} & \text{H—C—OH} \\
| & | \\
\text{HO}\overset{3}{-}\text{C—H} & \text{HO—C—H} \\
| & | \\
\text{H}\overset{4}{-}\text{C—OH} & \text{H—C—OH} \\
| & | \\
\text{H} & \text{H} \\
\text{L-erythrose} & \text{L-threose}
\end{array}$$

```
        H                         H
        |                         |
        C=O                       C=O
        |                         |
   HO—C—H                    H—C—OH
        |                         |
    H—C—OH                    H—C—OH
        |                         |
    H—C—OH                    H—C—OH
        |                         |
        H                         H

     D-threose                D-erythrose
```

Note the structural differences between L- and D-erythrose. There is a change in configuration for the number 3 as well as the number 2 carbon atom. This is true for all compounds with more than one asymmetric carbon atom.

At this point, the student may realize that the names and structures are accumulating at an alarming rate. Fortunately, most naturally occurring carbohydrates are of the D series.

There are a total of eight optical isomers for the aldopentoses, but only four in the D series.

13

```
        H                         H
        |                         |
        C=O                       C=O
        |                         |
    H—C—OH                    HO—C—H
        |                         |
    H—C—OH                    H—C—OH
        |                         |
    H—C—OH                    H—C—OH
        |                         |
    H—C—OH                    H—C—OH
        |                         |
        H                         H

     D-ribose                 D-arabinose


        H                         H
        |                         |
        C=O                       C=O
        |                         |
   HO—C—H                    H—C—OH
        |                         |
   HO—C—H                    HO—C—H
        |                         |
    H—C—OH                    H—C—OH
        |                         |
    H—C—OH                    H—C—OH
        |                         |
        H                         H

     D-lyxose                 D-xylose
```

Even though there are three asymmetric carbon atoms in the pentoses, there are only two variable asymmetric carbon atoms in the D series. Of these pentoses D-ribose is of the greater biologic importance.

Another pentose, the 2-deoxy derivative of ribose, represents a different series and is of considerable biologic importance.

14
$$\begin{array}{c} H \\ | \\ C=O \\ | \\ H-C-H \\ | \\ H-C-OH \\ | \\ H-C-OH \\ | \\ H-C-OH \\ | \\ H \end{array}$$
2-deoxy-D-ribose

There is no hydroxy group on the number 2 carbon, thus it is designated 2-deoxy. It, however, fulfills all requirements of a carbohydrate, and the relative arrangement of the groups bonded to the number 4 carbon places it in the D series.

The hexoses comprise the next higher homologs. One important example of the eight D-hexose isomers is D-glucose.

15
$$\begin{array}{c} H \\ | \\ C=O \\ | \\ H-C-OH \\ | \\ HO-C-H \\ | \\ H-C-OH \\ | \\ H-C-OH \\ | \\ H-C-OH \\ | \\ H \end{array}$$
D-glucose

The aldohexoses have four asymmetric carbons, but there are only three for consideration if the isomers are restricted to the D series. The number of isomers can be calculated by substitution in the following equation:

16   2^n = number of isomers
                     n = number of asymmetric carbon atoms

There are four asymmetric carbons in both the D and L series of the aldohexoses. Therefore, the total number of isomers would be 2^4 or 16. This equation can also be used to calculate the total number of isomers of the D series; thus, in the case of D-glucose, 2^3 or 8.

As previously stated, carbohydrates can exist in two distinct forms: those with an aldehyde group and those with a ketone group. Aldohexoses of primary biologic importance are D-glucose (**15**) and D-galactose (**17**).

17

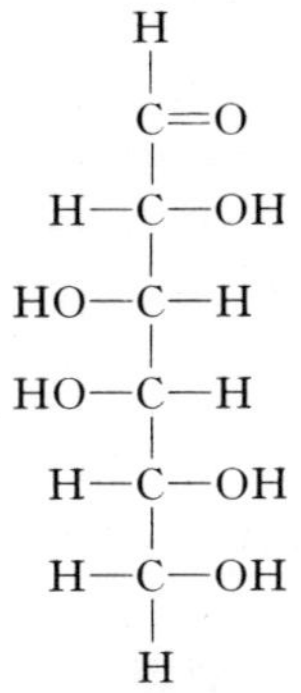

$$\begin{array}{c} H \\ | \\ C=O \\ | \\ H-C-OH \\ | \\ HO-C-H \\ | \\ HO-C-H \\ | \\ H-C-OH \\ | \\ H-C-OH \\ | \\ H \end{array}$$

D-galactose

The two ketoses of major interest are dihydroxyacetone (**5**) and fructose.

18

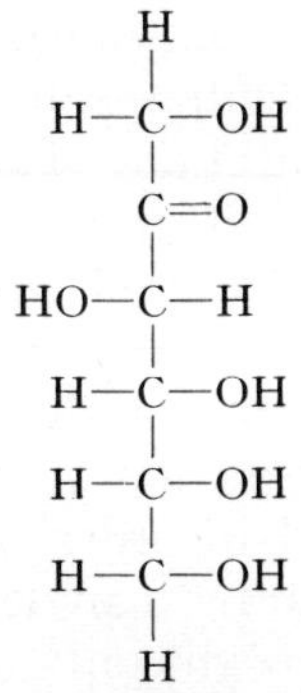

$$\begin{array}{c} H \\ | \\ H-C-OH \\ | \\ C=O \\ | \\ HO-C-H \\ | \\ H-C-OH \\ | \\ H-C-OH \\ | \\ H-C-OH \\ | \\ H \end{array}$$

D-fructose

REACTIONS

Carbohydrates have a number of interesting reactions because they possess many functional groups. Certainly one of the earliest reactions, and undoubtedly the most important to mankind, is fermentation to produce ethanol. Anaerobic fermentation leads to the production of ethanol.

19
$$C_6H_{12}O_6 \xrightarrow[\text{fermentation}]{\text{anaerobic}} 2\,C_2H_5OH + 2\,CO_2$$

Aerobic fermentation yields acetic acid and certain other organic acids. The production of vinegar by aerobic fermentation is of commercial importance.

20
$$C_6H_{12}O_6 + O_2 \xrightarrow[\text{fermentation}]{\text{aerobic}} 2\,CH_3COOH + 2\,CO_2 + 2\,H_2$$

The above reactions are schematic at best in that there are a number of other products produced, and the yields are not quantitative as written. Both types of fermentation, depending upon the organism and conditions, lead to a number of commercially useful products.

Carbohydrates such as glucose and fructose react readily with a number of metals in mildly alkaline solution to reduce the metal and yield a number of oxidation products.

21
$$\begin{array}{c}CHO\\|\\(HCOH)_4\\|\\CH_2OH\end{array} + 2\,Cu(OH)_2 \xrightarrow{\Delta T} Cu_2O + 2\,H_2O + \begin{array}{c}COOH\\|\\(HCOH)_4\\|\\CH_2OH\end{array}$$

This reaction is not as simple as shown.

When heated in mild alkaline solutions, carbohydrates undergo what is often termed fragmentation. The carbohydrates are degraded to yield a number of smaller reactive molecules. These molecules react with metals leading to a much greater reduction of the metal than might be expected. Timing is very essential in the reaction because the number of reducing products formed is time dependent. Whenever the reaction is used for quantitation the length of time must be precisely controlled in order to obtain meaningful results. The oxidation of carbohydrates can lead to several products (**22**). The conversion of glucose to gluconic acid by the oxidation of the aldehyde group to a carboxyl group can be accomplished in the laboratory without difficulty. However, under laboratory conditions, oxidation of glucose to glucosiduronic acid is difficult because it involves the oxidation of a primary alcohol group in a molecule which contains an aldehyde group.

This process is readily accomplished in biologic systems. Further oxidation of either gluconic or glucosiduronic acid leads to the dicarboxylic oxidation product glucaric acid. The oxidation of the aldohexoses to the corresponding dicarboxylic acid is readily accomplished in the laboratory. Similarly, galactose can undergo oxidation to form a dicarboxylic acid, mucic acid. This reaction is often important in the identification of galactose.

22

$$
\begin{array}{c}
\text{CHO} \\
| \\
\text{H—C—OH} \\
| \\
\text{HO—C—H} \\
| \\
\text{H—C—OH} \\
| \\
\text{H—C—OH} \\
| \\
\text{H—C—OH} \\
| \\
\text{H}
\end{array}
\quad\xrightarrow{\text{oxidation}}\quad
\begin{array}{c}
\text{COOH} \\
| \\
\text{H—C—OH} \\
| \\
\text{HO—C—H} \\
| \\
\text{H—C—OH} \\
| \\
\text{H—C—OH} \\
| \\
\text{H—C—OH} \\
| \\
\text{H}
\end{array}
$$

D-glucose → D-gluconic acid

↓ oxidation   ↓ oxidation

$$
\begin{array}{c}
\text{CHO} \\
| \\
\text{H—C—OH} \\
| \\
\text{HO—C—H} \\
| \\
\text{H—C—OH} \\
| \\
\text{H—C—OH} \\
| \\
\text{COOH}
\end{array}
\quad\xrightarrow{\text{oxidation}}\quad
\begin{array}{c}
\text{COOH} \\
| \\
\text{H—C—OH} \\
| \\
\text{HO—C—H} \\
| \\
\text{H—C—OH} \\
| \\
\text{H—C—OH} \\
| \\
\text{COOH}
\end{array}
$$

D-glucosiduronic acid   D-glucaric acid

Another reaction which has considerable importance in the identification of sugars is the reaction of the sugars with phenylhydrazine.

23

$$
\begin{array}{c}
\text{CHO} \\
| \\
\text{H—C—OH} \\
| \\
\text{HO—C—H} \\
| \\
\text{H—C—OH} \\
| \\
\text{H—C—OH} \\
| \\
\text{CH}_2\text{OH}
\end{array}
+ \text{NH}_2\text{NH—C}_6\text{H}_5
\;\xrightarrow{-\text{H}_2\text{O}}\;
\begin{array}{c}
\text{H} \\
| \\
\text{C=N—NH—C}_6\text{H}_5 \\
| \\
\text{H—C—OH} \\
| \\
\text{HO—C—H} \\
| \\
\text{H—C—OH} \\
| \\
\text{H—C—OH} \\
| \\
\text{CH}_2\text{OH}
\end{array}
\;\xrightarrow{\text{NH}_2\text{NH—C}_6\text{H}_5}\;
$$

D-glucose   phenylhydrazine   glucose phenylhydrazone

$$\begin{bmatrix} \text{H} \\ | \\ \text{C}=\text{N}-\text{NH}-\text{C}_6\text{H}_5 \\ | \\ \text{C}=\text{O} \\ | \\ \text{HO}-\text{C}-\text{H} \\ | \\ \text{H}-\text{C}-\text{OH} \\ | \\ \text{H}-\text{C}-\text{OH} \\ | \\ \text{CH}_2\text{OH} \end{bmatrix} + \text{NH}_2-\text{C}_6\text{H}_5 + \text{NH}_3 \xrightarrow[\text{NH}_2\text{NH}-\text{C}_6\text{H}_5]{-\text{H}_2\text{O}} \begin{array}{c} \text{H} \\ | \\ \text{C}=\text{N}-\text{NH}-\text{C}_6\text{H}_5 \\ | \\ \text{C}=\text{N}-\text{NH}-\text{C}_6\text{H}_5 \\ | \\ \text{HO}-\text{C}-\text{H} \\ | \\ \text{H}-\text{C}-\text{OH} \\ | \\ \text{H}-\text{C}-\text{OH} \\ | \\ \text{CH}_2\text{OH} \end{array}$$

aniline

glucosazone

The first mole of phenylhydrazine reacts with glucose to produce the hydrazone. The hydrazone is then oxidized in some way such that a second molecule of phenylhydrazine attaches to yield the osazone. The mechanism of this reaction is disputed by some investigators; therefore, part of the reaction is enclosed in brackets. Regardless of what the mechanism is, three moles of phenylhydrazine are required to produce one mole of osazone product.

Fructose reacts with phenylhydrazine in a similar manner.

24

$$\begin{array}{c} \text{CH}_2\text{OH} \\ | \\ \text{C}=\text{O} \\ | \\ \text{HO}-\text{C}-\text{H} \\ | \\ \text{H}-\text{C}-\text{OH} \\ | \\ \text{H}-\text{C}-\text{OH} \\ | \\ \text{CH}_2\text{OH} \end{array} + \text{NH}_2\text{NH}-\text{C}_6\text{H}_5 \xrightarrow{-\text{H}_2\text{O}} \begin{array}{c} \text{CH}_2\text{OH} \\ | \\ \text{C}=\text{N}-\text{NH}-\text{C}_6\text{H}_5 \\ | \\ \text{HO}-\text{C}-\text{H} \\ | \\ \text{H}-\text{C}-\text{OH} \\ | \\ \text{H}-\text{C}-\text{OH} \\ | \\ \text{CH}_2\text{OH} \end{array} \xrightarrow{\text{NH}_2\text{NH}-\text{C}_6\text{H}_5}$$

D-fructose   phenylhydrazine   fructose phenylhydrazone

$$\begin{bmatrix} \text{H} \\ | \\ \text{C}=\text{O} \\ | \\ \text{C}=\text{N}-\text{NH}-\text{C}_6\text{H}_5 \\ | \\ \text{HO}-\text{C}-\text{H} \\ | \\ \text{H}-\text{C}-\text{OH} \\ | \\ \text{H}-\text{C}-\text{OH} \\ | \\ \text{CH}_2\text{OH} \end{bmatrix} + \text{C}_6\text{H}_5-\text{NH}_2 + \text{NH}_3 \xrightarrow[\text{NH}_2\text{NH}-\text{C}_6\text{H}_5]{-\text{H}_2\text{O}} \begin{array}{c} \text{H} \\ | \\ \text{C}=\text{N}-\text{NH}-\text{C}_6\text{H}_5 \\ | \\ \text{C}=\text{N}-\text{NH}-\text{C}_6\text{H}_5 \\ | \\ \text{HO}-\text{C}-\text{H} \\ | \\ \text{H}-\text{C}-\text{OH} \\ | \\ \text{H}-\text{C}-\text{OH} \\ | \\ \text{CH}_2\text{OH} \end{array}$$

fructosazone (glucosazone)

Again the reaction mechanism is not agreed upon, but presumably the process is similar to that for glucose. Note that the product formed from fructose is the same as that produced by glucose. The differences in carbons 1 and 2 of the glucose and fructose molecules are eliminated during the formation of osazones. Since carbons 3, 4, 5, and 6 of glucose and fructose are identical, so are the osazones.

The importance of osazones is they allow the formation of crystalline derivatives of the sugars. Carbohydrates are very soluble in aqueous solvents but quite insoluble in most organic solvents. It is very difficult to crystallize them from most solvent systems, and thus it is essentially impossible to crystallize carbohydrates in a pure form from biologic systems. Osazones provide a means of isolating carbohydrates from biologic sources, and they can be used as aids in the identification of sugars. Unfortunately the melting points of the osazones of the biologically important sugars are very similar and usually identification cannot be made on this basis. However, the crystal structures of the osazones are sufficiently different so that when they are compared microscopically with osazones of known sugars identification can usually be made.

The reaction of aldoses with o-toluidine to produce a colored compound is very important in clinical chemistry.

25

$$\begin{array}{c} H \\ C=O \\ | \\ H-C-OH \\ | \\ HO-C-H \\ | \\ H-C-OH \\ | \\ H-C-OH \\ | \\ CH_2OH \end{array} + NH_2\text{-}C_6H_4\text{-}CH_3 \longrightarrow \begin{array}{c} H \\ C=N\text{-}C_6H_4\text{-}CH_3 \\ | \\ H-C-OH \\ | \\ HO-C-H \\ | \\ H-C-OH \\ | \\ H-C-OH \\ | \\ CH_2OH \end{array} + H_2O$$

D-glucose   o-toluidine

This reaction is the basis of what is probably the best method for measuring glucose in serum, spinal fluid, and urine.

Under the influence of hot mineral acids, the sugars undergo dehydration to yield heterocyclic aldehydes. The importance of these reactions is that they are a part of several procedures used to quantitate the sugars.

26 A

$$\begin{array}{c} H \\ | \\ C=O \\ | \\ H-C-OH \\ | \\ H-C-OH \\ | \\ H-C-OH \\ | \\ H-C-OH \\ | \\ H \end{array} \xrightarrow[\text{acid}]{-3\,H_2O} \text{furfural}$$

D-ribose

26 B

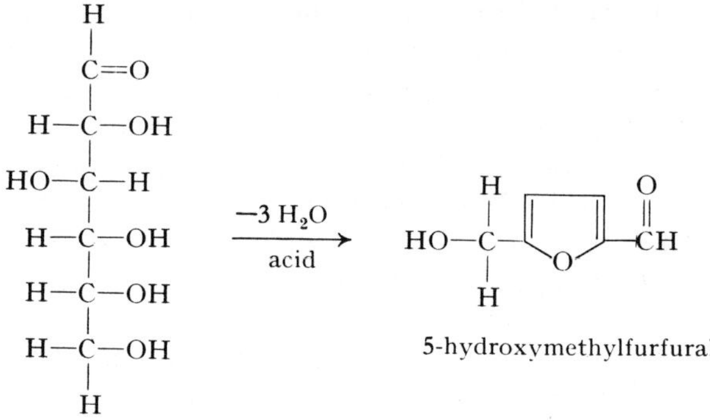

D-glucose (H, C=O, H–C–OH, HO–C–H, H–C–OH, H–C–OH, H–C–OH, H) →[−3 H₂O, acid]→ 5-hydroxymethylfurfural (HOCH₂–furan–CHO)

The furfurals condense with a number of organic compounds to yield colored derivatives which can be used in colorimetric procedures. For example, 5-hydroxymethylfurfural reacts with anthrone to yield a colored derivative.

27

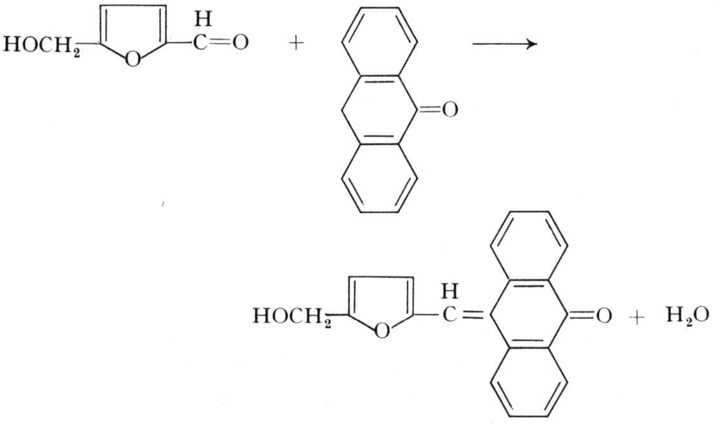

HOCH₂–furan–CHO + anthrone (with C=O) ⟶ HOCH₂–furan–CH=C(anthracenylidene)=O + H₂O

blue-green

This reaction is a basis of a widely used procedure for the measurement of carbohydrates in biologic samples.

HEMIACETAL STRUCTURES

Carbonyl groups are extremely active in aqueous solvents and readily react with water to yield hydrated forms. An example of this for an aldehyde is

28
$$CH_3\overset{O}{\overset{\|}{C}}-H + H_2O \rightleftharpoons CH_3-\underset{OH}{\overset{OH}{\underset{|}{\overset{|}{C}}}}-H$$

        acetaldehyde   hydrated acetaldehyde

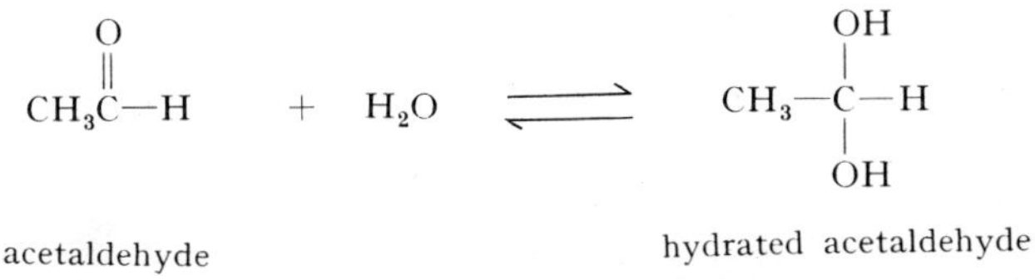

A similar reaction occurs with ketones. Analogous reactions occur with alcohols.

29
$$CH_3-\overset{O}{\overset{\|}{C}}-H + HOCH_3 \longrightarrow CH_3-\underset{OCH_3}{\overset{OH}{\underset{|}{\overset{|}{C}}}}-H$$

    acetaldehyde   methanol   hemiacetal

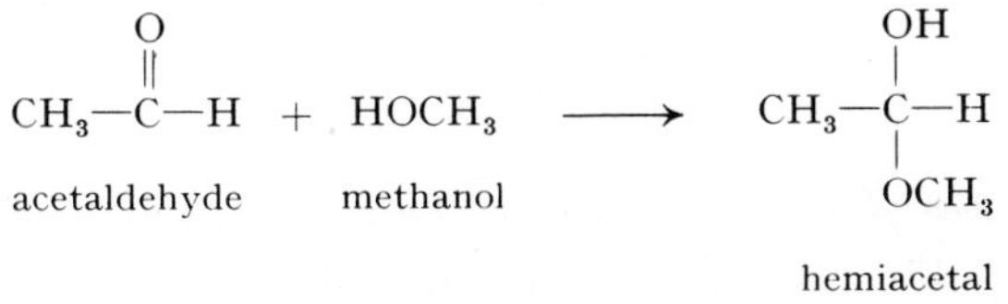

Acetaldehyde reacts with methanol to form an addition product called a hemiacetal. If the structure of the glucose is considered in terms of these functional groups, it is apparent that the molecule contains an aldehyde group and alcohol groups. It is perhaps then not surprising to realize that glucose can form hemiacetals. What may be surprising, however, is that it forms an intramolecular hemiacetal.

30

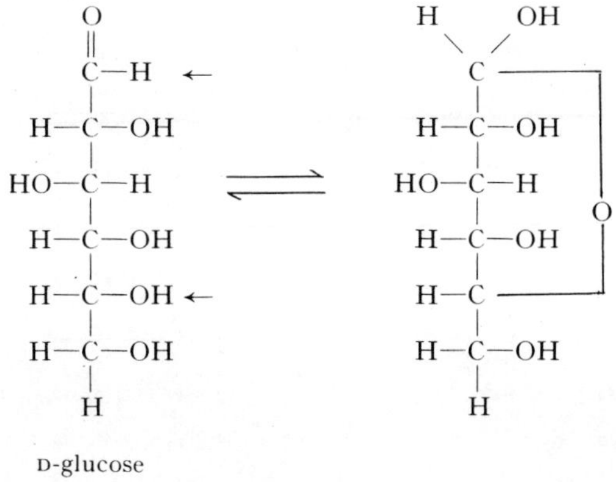

$$\begin{array}{c} \overset{O}{\overset{\|}{C}}-H \leftarrow \\ H-C-OH \\ HO-C-H \\ H-C-OH \\ H-C-OH \leftarrow \\ H-C-OH \\ H \end{array} \rightleftharpoons \begin{array}{c} H\quad OH \\ \diagdown\diagup \\ C- \\ H-C-OH \\ HO-C-H \\ H-C-OH \\ H-C- \\ H-C-OH \\ H \end{array} \bigg] O$$

D-glucose

The intramolecular hemiacetal shown in **30** has the oxygen bridge between carbons 1 and 5. This is probably the most stable configuration of the molecule; however, it can exist with the oxygen bridge to other hydroxy groups. A similar reaction occurs with fructose.

31

```
        CH₂OH              CH₂OH
         |                  |
         C=O             HO—C ──┐
         |                  |   │
     HO—C—H             HO—C—H  │
         |                  |   │
      H—C—OH             H—C—OH O
         |                  |   │
      H—C—OH             H—C ───┘
         |                  |
      H—C—OH             H—C—OH
         |                  |
         H                  H
```

⇌

D-fructose

The oxygen bridge here is shown extending to carbons 2 and 5. This is presumably the most favorable configuration for fructose, but it should be pointed out that the oxygen bridge can also occur between carbon 2 and other carbons.

Examination of the hemiacetal form of D-glucose (**30**) reveals that the molecule now contains five asymmetric carbon atoms, because carbon 1 is now joined to four different groups, and thus, it too is asymmetric. This means then that for the aldohexoses there are a total of 32 isomers, or of the D series there is a total of 16 isomers, or indeed, for D-glucose there are 2 isomers. These are shown in **32**.

32

```
      HO   H              H    OH
        \ /                \  /
         C ──┐              C ──┐
         |   │              |   │
      H—C—OH │           H—C—OH │
         |   │              |   │
     HO—C—H  O          HO—C—H  O
         |   │              |   │
      H—C—OH │           H—C—OH │
         |   │              |   │
      H—C ───┘           H—C ───┘
         |                  |
      H—C—OH             H—C—OH
         |                  |
         H                  H
```

β-D-glucose   α-D-glucose

These sterioisomers involving the number 1 carbon are different in nature than those involving other carbons. The α and β isomers of glucose do indeed show different properties, particularly optical properties, but they are different than isomers involving hydroxyl groups on other carbons because they are unstable configurations. As seen in (**33**), the β and α forms are in equilibrium in solution with the free aldehyde form.

33

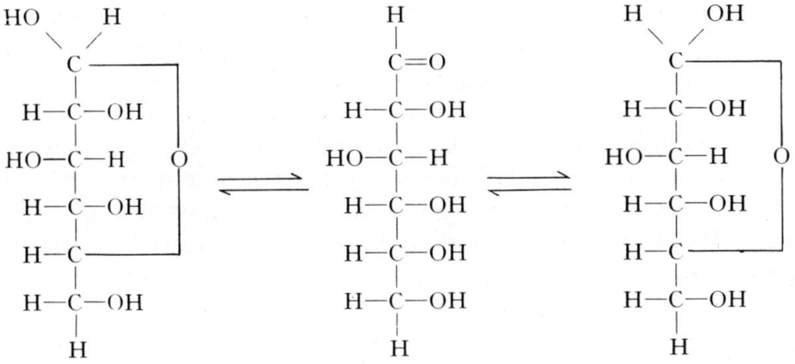

β-D-glucose ⇌ D-glucose ⇌ α-D-glucose

Glucose in solution shows very little of the properties expected of aldehydes. At equilibrium glucose is mainly in the form of a hemiacetal and very little exists as a free aldehyde. In reactions involving the carbonyl carbons, the equilibrium is readily displaced in order to provide the reactive aldehyde form.

An examination of the hemiacetals as they have been written previously — that is **30, 31, 32, 33** — shows that the oxygen bridge connecting the carbonyl carbons with the hydroxyl groups is of an unusual nature. It is a long bond and has a couple of right angles in it. This sort of bonding certainly occurs only on paper. A more realistic representation of the hemiacetal forms can be accomplished by drawing them as more conventional heterocyclic ring forms. A heterocyclic ring which closely resembles the hemiacetal form of glucose is the structure of pyran.

34

pyran

Glucose is represented by this technique as

35

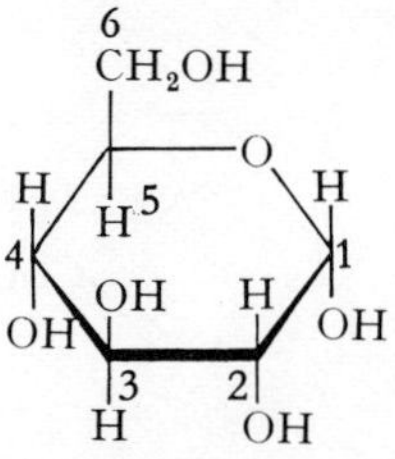

α-D-glucose
α-D-glucopyranose

In order to draw glucose in the pyran form certain conventions are observed. As shown in **35**, the number 1 carbon occurs about 3 o'clock and the numbering proceeds in a clockwise fashion around the ring. If a hydroxyl group is on the right side of the chain when the molecule is drawn in a straight chain form (**33**), it is drawn below the plane of the ring. If it occurs on the left side of the chain as for carbon 3 (**33**), the hydroxyl group is written above the plane of the ring. The front edge of the pyran ring is indicated by the heavy line. However, in practice the structure is not usually drawn as shown in **35**, but rather as

36

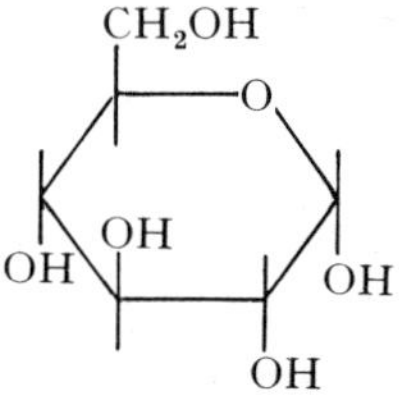

$$\text{CH}_2\text{OH}$$

(structure with OH groups)

α-D-glucopyranose

As you will note, the front edge of the ring is not darkened to show that it projects forward and downward; also, the hydrogens are omitted. This is done for convenience. It is not unlike the same lax methods that are used in representing benzene and other ring forms. A new name for glucose, glucopyranose, indicates that the molecule occurs in the heterocyclic pyran ring. The α indicates that the hydroxy group is in the α position for the number 1 carbon and that the molecule belongs to the D series as indicated by the small capital D. The general configuration of the molecule is that of glucose. The pyran portion of the name, as mentioned, refers to the heterocyclic structure, and the ose ending (as always) indicates that it is a carbohydrate.

Similarly, fructose can be written in better form as a derivative of furan.

37

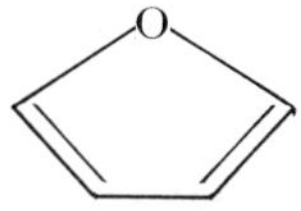

furan

Fructose also forms β and α sterioisomers.

38

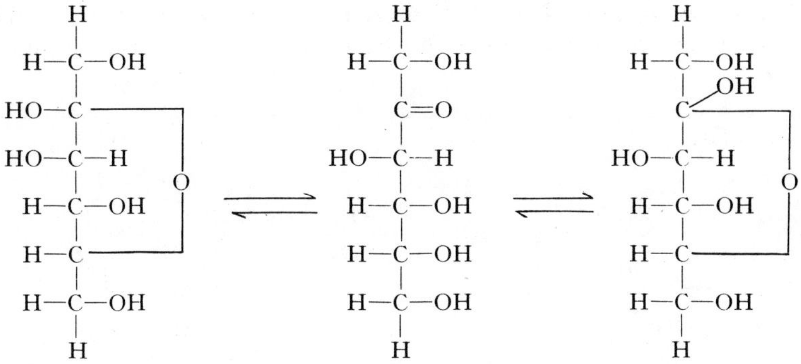

$$\beta\text{-D-fructose} \rightleftharpoons \text{D-fructose} \rightleftharpoons \alpha\text{-D-fructose}$$

Fructose in the furan ring form follows the conventions for glucose.

39

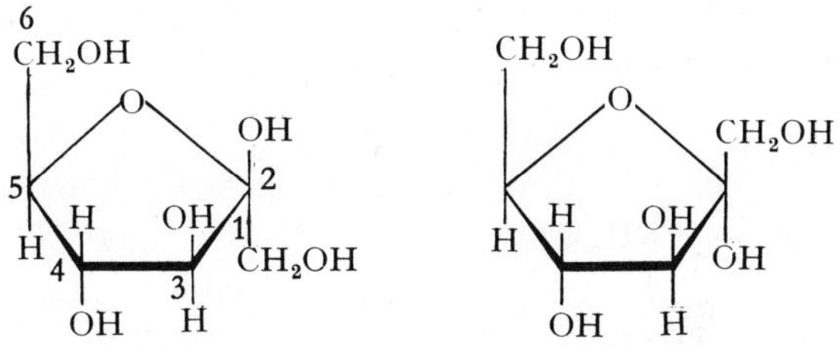

β-D-fructofuranose   α-D-fructofuranose

Note, however, that the number 1 carbon in the furan form is outside the ring.

The student may well believe at this point that the representation of the sugars in the pyran and furan ring forms only unnecessarily complicates things. However, when the structures of the disaccharides are considered, the student may agree that representation in the pyran and furan forms actually makes things easier.

CARBOHYDRATE DERIVATIVES

Even though the carbonyl carbon of the sugars is in the hemiacetal form, it still retains great reactivity. An example of this is when glucose reacts with an alcohol in the presence of an acid to form an ether derivative with the hydroxyl group on the number 1 carbon.

40

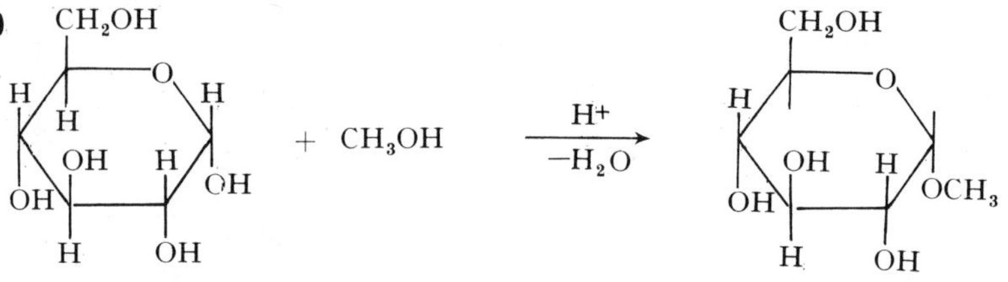

α-D-glucopyranose + CH_3OH $\xrightarrow{H^+, -H_2O}$ methyl-α-D-glucopyranoside

This type of derivative is known as a glycoside.

The carbohydrates, particularly in biologic reactions, can form a number of other derivatives, such as the reaction of fructose with phosphoric acid to form fructose-1-phosphate.

41

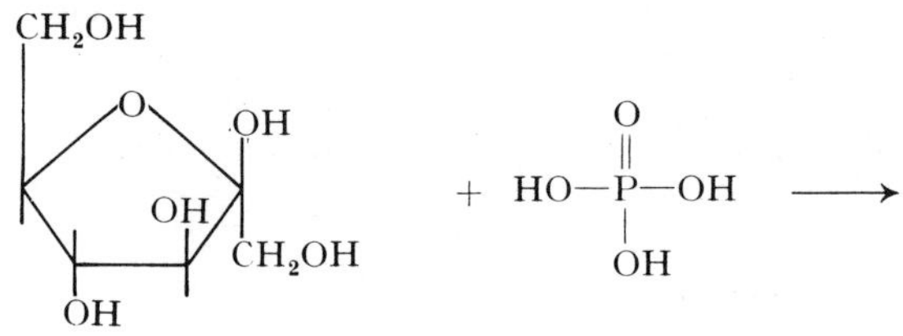

β-D-fructofuranose + $HO-P(=O)(OH)-OH$ ⟶

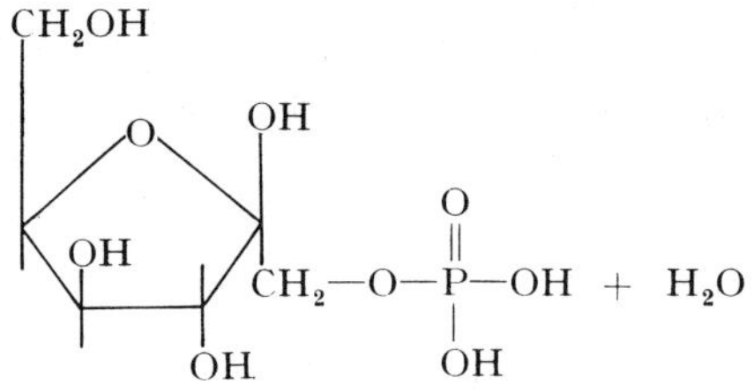

β-D-fructofuranose-1-phosphate
(fructose-1-phosphate) + H_2O

The phosphate group has not reacted with the carbonyl carbon, but rather with the number 1 primary alcohol group. Carbohydrates can also form derivatives with the primary alcohol group on the other end of the chain. Fructose can react with phosphoric acid to yield fructose-6-phosphate.

42

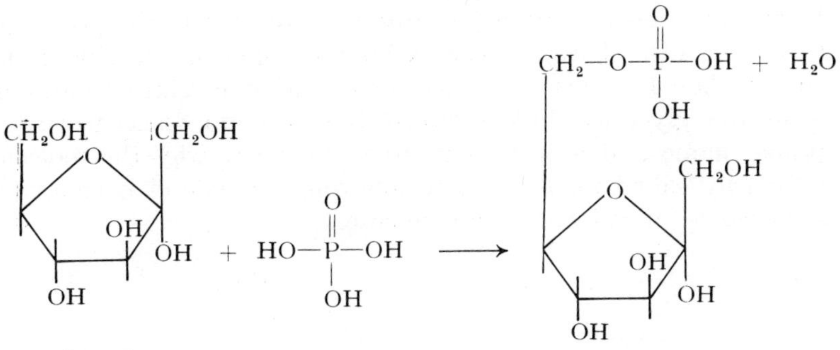

[α-D-fructofuranose] + HO—P(=O)(OH)—OH ⟶ [fructose-6-phosphate (6-phosphofructose)] + CH$_2$—O—P(=O)(OH)—OH + H$_2$O

α-D-fructofuranose

fructose-6-phosphate
(6-phosphofructose)

These derivatives of the primary alcohol group are difficult to make in the laboratory but are readily produced in biologic systems. Later on the student will see that in biologic systems a very common intermediate of metabolism is fructose-1,6-diphosphate.

DISACCHARIDES

Up until this point, only carbohydrates which have carbons joined to each other in a continuous carbon chain have been considered. These are called the monosaccharides. There are carbohydrates consisting of two monosaccharides which are connected by an acetal or a glycoside link.

43

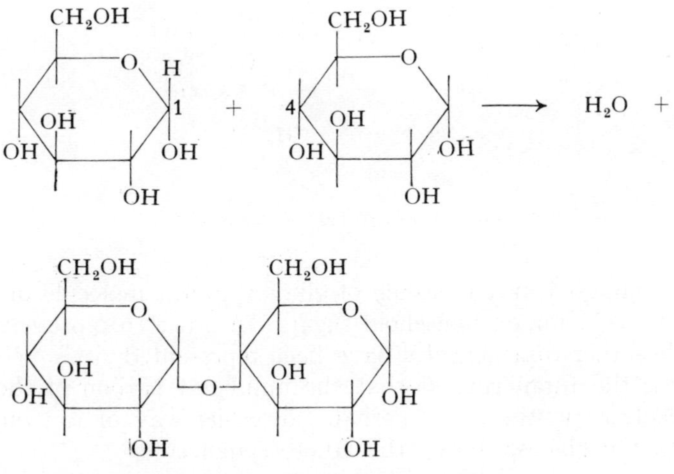

maltose
4-(α-D-glucopyranosyl)-α-D-glucopyanose

In this reaction two molecules of glucose react with the loss of water to form a disaccharide. The carbonyl carbon of one glucose molecule, carbon 1, reacts with carbon 4 of the other glucose molecule to form the glycoside link between the two monosaccharides. The proper name makes it obvious to the student why the molecule is ordinarily called maltose. The following compound is composed of a molecule of galactose and glucose.

44

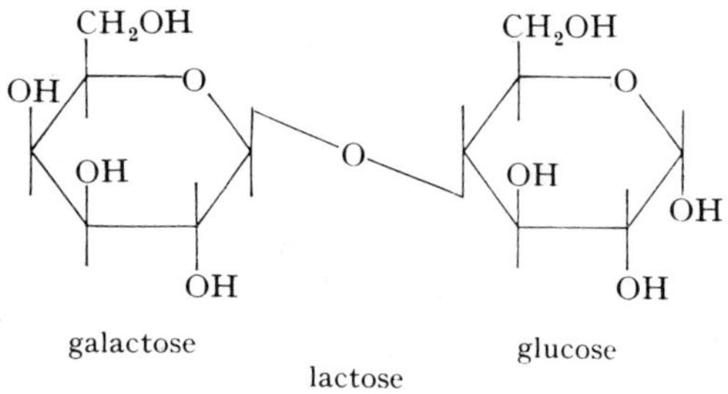

galactose — glucose
lactose

4-(β-D-galactopyranosyl)-α-D-glucopyanose

Note that the link between the galactose and the glucose involves a β configuration of the galactosyl group. Again, it should be obvious why the molecule is called lactose rather than its official name.

Perhaps the organic molecule with the greatest economic importance in all history is the disaccharide sucrose.

45

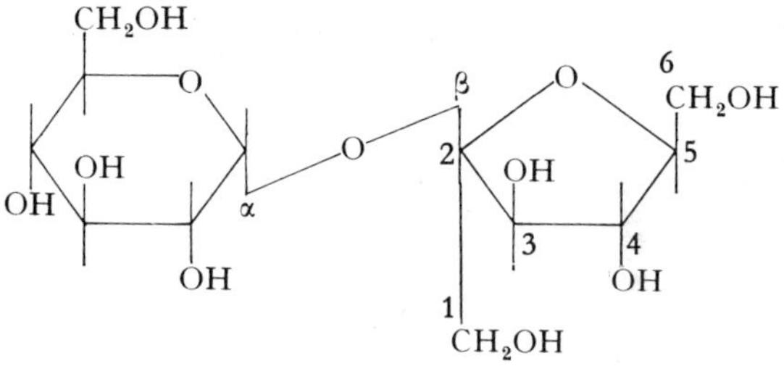

sucrose

α-D-glucopyranosyl-β-D-fructofuranoside

It is composed of a molecule of glucose and a molecule of fructose. Sucrose is common household sugar. In order to represent fructose as the other disaccharides have been represented, it is necessary to reverse the furan ring so that the number 1 carbon of the fructose molecule is on the left. Perhaps an easier way of drawing sucrose shows the glucose above the fructose molecule.

46

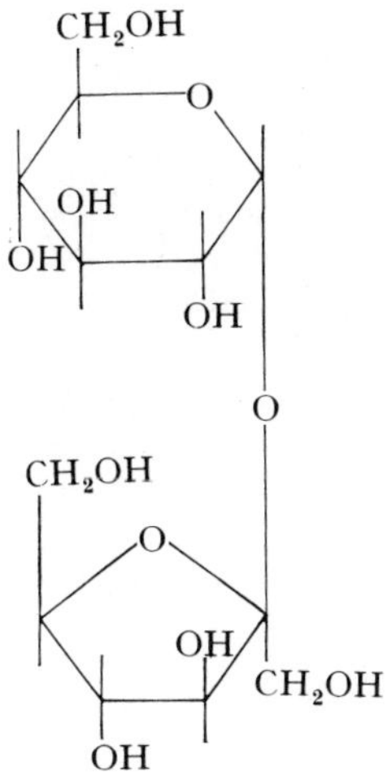

sucrose

Examination of the structure of both maltose (**43**) and lactose (**44**) reveals that the monosaccharides located on the right, as written, retain a carbonyl group in the hemiacetal form. Also, each molecule located on the left, as written, has a carbonyl group involved in a glycoside link. Glycoside links are resistant to alkaline hydrolysis but are readily hydrolyzed by acid solutions. Thus, in reducing or other reactions involving carbonyl carbons, the molecules on the left of the disaccharide molecules will be unreactive. Those on the right with a free carbonyl group will have the expected reactivities. Sucrose, however, is different (**46**). The carbonyl carbons of both the monosaccharides (fructose and glucose) are involved in a glycoside link, therefore preventing carbonyl reactivity. For sucrose to show carbonyl reactivities like other disaccharides, it must first be hydrolyzed by acid.

If the student still questions the wisdom of representing the carbohydrates in the heterocyclic ring forms, he might try drawing some of the disaccharides with the straight chain formulas.

POLYSACCHARIDES

Monosaccharides can combine by glycoside links to form trisaccharides, tetrasaccharides, and so on. These generally are not of great biologic importance. However, the larger polymers of disaccharides are of considerable interest. A polymer of glucose units linked 1 → 4, as in maltose, is called amylose.

47

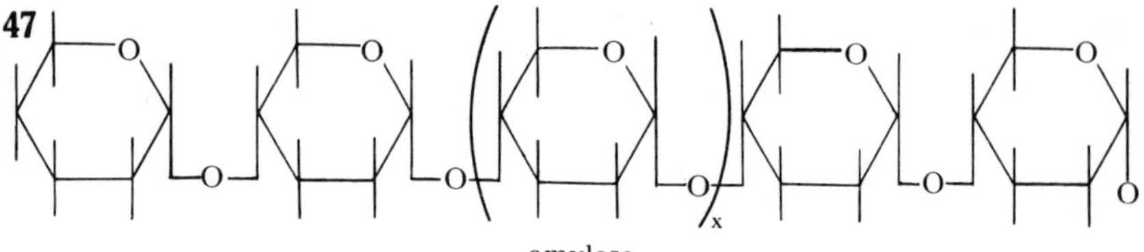

amylose

Most monosaccharide polymers, particularly glucose, occur as branched structures.

48

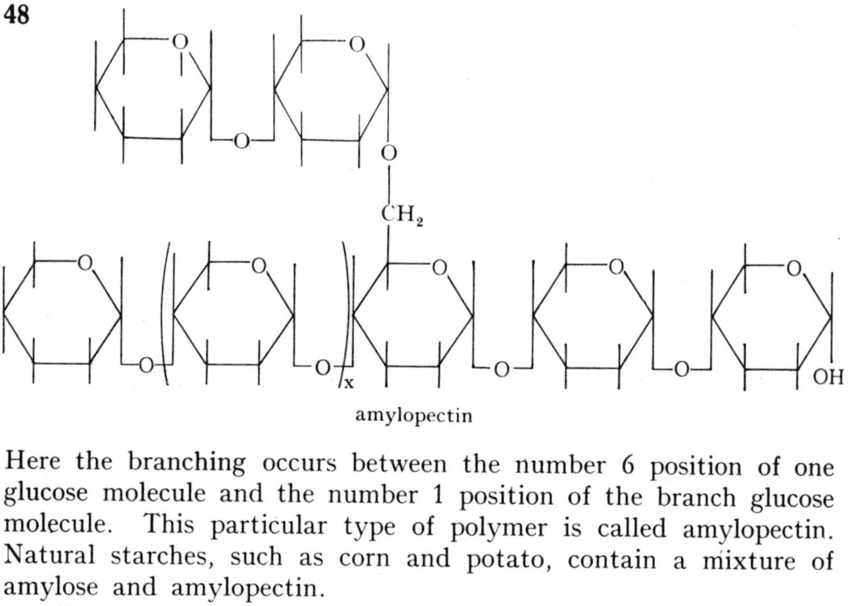

amylopectin

Here the branching occurs between the number 6 position of one glucose molecule and the number 1 position of the branch glucose molecule. This particular type of polymer is called amylopectin. Natural starches, such as corn and potato, contain a mixture of amylose and amylopectin.

The amylose and amylopectin polymers both have only one monosaccharide group which has a free carbonyl group. These carbonyl groups are located at the right terminals of the molecules. Due to the high molecular weight of these polymers with only one reactive group per molecule, the carbonyl properties are insignificant.

IMPORTANCE OF CARBOHYDRATES

To this point only the structures and some of the chemical properties of carbohydrates have been presented. Carbohydrates are important in the metabolism of biologic systems, as well as in commerce. Glyceraldehyde and dihydroxyacetone are important intermediates in carbohydrate metabolism. Other carbohydrates of predominant interest are the pentoses, D-ribose and 2-deoxy-D-ribose. The hexoses —glucose, galactose, and fructose— are of primary interest. Glucose is considered the common intermediate of carbohydrate metabolism, is a normal constituent of blood, and is a prime source of energy for most living systems. All other carbohydrates are converted to glucose in normal metabolism. Galactose and fructose do not usually occur in biologic systems in large concentration in the free state. Perhaps it would be of interest to know that fructose is the sweetest known sugar.

Of the disaccharides, maltose does not normally occur free in nature. It is, however, a product of amylase hydrolysis of polysaccharides. Lactose is a common constituent of mammalian milk and in the past has been referred to as milk sugar. Sucrose, of course, is a very common sugar. It is produced and accumulated in large amounts in a number of plants, particularly sugar cane and sugar beet.

As indicated previously, common starches are a mixture of the polysaccharides, amylose, and amylopectin. Another polysaccharide molecule of importance is cellulose, a polymer which constitutes a large part of wood and cotton. It consists of glucose units linked very much like amylose, except the linkage is in the β form. Heparin, a naturally occurring anticoagulant, and agar-agar, a product obtained from seaweed and extensively used in growth media for microorganisms, are also polysaccharides.

CHAPTER 2

LIPIDS

Very simply, lipids are a confusing conglomerate. This situation has evolved mainly because the qualifications for being a lipid are not very discriminating. The following list of requirements substantiates this statement:

A lipid must be:

1. From a biologic system
2. Soluble in lipid solvents (diethyl ether, benzene, toluene, chloroform, acetone, and ethanol)
3. Insoluble in water

With loose entrance requirements, it is obvious that this group of compounds would contain many heterogenous substances. The various classification schemes devised by workers in an effort to clarify the subject has probably only made the matter worse. To be conventional, lipids will be considered here according to a very simple classification scheme.

NEUTRAL LIPIDS

TRIGLYCERIDES

The first class of compounds to be considered is the neutral lipids; these consist of esters of fatty acids with various alcohols. Fatty acids are monocarboxylic straight-chain acids. Most of the naturally occurring fatty acids contain an even number of carbon atoms. The reason for this will become apparent when fatty acid metabolism is discussed. Refer to Table 2-1 for a listing of the fatty acids.

TABLE 2-1. FATTY ACIDS

Name		Formulas
Common	Systemic	
Acetic	—	CH_3COOH
Butyric	—	$CH_3(CH_2)_2COOH$
Caproic	Hexanoic	$CH_3(CH_2)_4COOH$
Caprylic	Octanoic	$CH_3(CH_2)_6COOH$
Capric	Decanoic	$CH_3(CH_2)_8COOH$
Lauric	Dodecanoic	$CH_3(CH_2)_{10}COOH$
Myristic	Tetradecanoic	$CH_3(CH_2)_{12}COOH$
Palmitic*	Hexadecanoic	$CH_3(CH_2)_{14}COOH$
Stearic*	Octadecanoic	$CH_3(CH_2)_{16}COOH$

* Palmitic and stearic acids occur most often in neutral lipids.

The most common alcohol in the neutral lipids is glycerol.

49

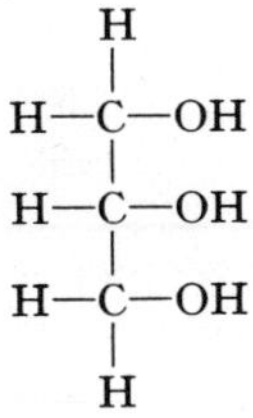

$$\begin{array}{c} H \\ | \\ H-C-OH \\ | \\ H-C-OH \\ | \\ H-C-OH \\ | \\ H \end{array}$$

glycerol
(glycerin)

Most of the lipids isolated from fat depots are triesters of glycerin with fatty acids.

50

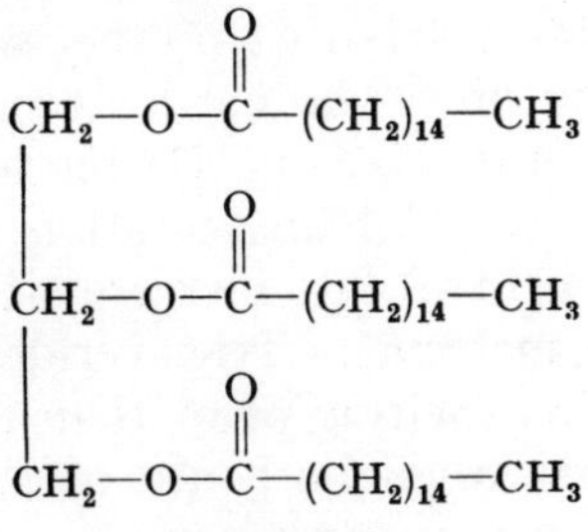

$$\begin{array}{l} CH_2-O-\overset{\overset{\displaystyle O}{\|}}{C}-(CH_2)_{14}-CH_3 \\ | \\ CH_2-O-\overset{\overset{\displaystyle O}{\|}}{C}-(CH_2)_{14}-CH_3 \\ | \\ CH_2-O-\overset{\overset{\displaystyle O}{\|}}{C}-(CH_2)_{14}-CH_3 \end{array}$$

tripalmitin

Unfortunately most of the triesters of glycerin, called triglycerides, do not contain just one fatty acid; usually they contain three different fatty acids. For example, a triglyceride can consist of esters with myristic acid, palmitic acid, and stearic acid all in the same molecule.

51

$$\begin{array}{l} CH_2-O-\overset{O}{\overset{\parallel}{C}}-(CH_2)_{12}-CH_3 \\ CH_2-O-\overset{O}{\overset{\parallel}{C}}-(CH_2)_{14}-CH_3 \\ CH_2-O-\overset{O}{\overset{\parallel}{C}}-(CH_2)_{16}-CH_3 \end{array}$$

myristopalmitostearin

In addition to the saturated fatty acids in Table 2-1, many fats contain unsaturated fatty acids such as oleic acid, a derivative of stearic acid.

52

$$\begin{array}{c} CH_3-(CH_2)_7-CH \\ \overset{O}{\overset{\parallel}{}} \qquad \qquad \parallel \\ HO-C-(CH_2)_7-CH \end{array}$$

oleic acid

Note the configuration. The geometric isomerism around the double bond is in the *cis* configuration. The student should recall this type of isomerism from earlier studies.

Properties of Triglycerides

Neutral lipids from different sources have different properties. For instance, hard fats, such as beef tallow, have higher melting points than triglycerides isolated from other sources, such as linseed, peanut, and cottonseed oil. The hardness or the melting point of a neutral fat depends on two factors: (1) the length of the fatty acid chains, and (2) the amount of unsaturation in the molecule. The melting points of the neutral fats increase with the chain length of the fatty acids in the molecule. Triglycerides containing saturated fatty acids have a higher melting point than triglycerides containing unsaturated fatty acids when both are of the same chain length. Neutral lipids with lower melting points, or softer fats, are often called oils. The term oil, however, should not be associated solely with the neutral fats or triglycerides since it is also used for mixtures of hydrocarbons such as motor oil, mineral oil, and so on. These oils are in no way chemically related to the fats of triglycerides.

Purified triglycerides have no flavor. Actually, the flavor associated with fat is due to impurities. In the case of butter, the characteristic flavor is caused by products formed by bacterial flora through the fermentation of carbohydrates.

Reactions of Triglycerides, Glycerol, and Fatty Acids

When the triglycerides are hydrolyzed with an alkali, the ester linkage is split, yielding the alcohol (glycerin) and salts of the fatty acids.

53

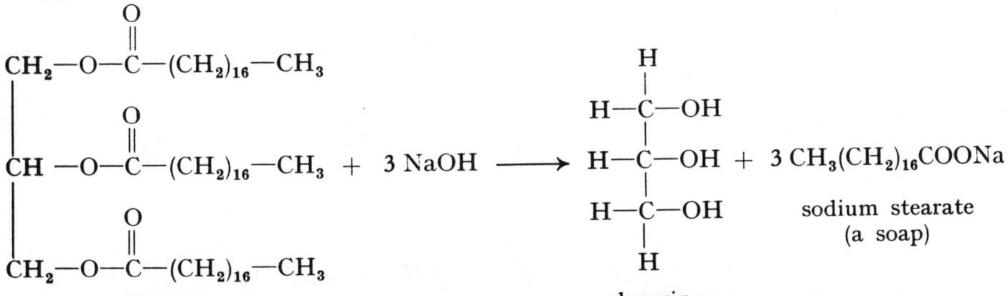

$$\begin{array}{l} CH_2-O-\overset{O}{\overset{\|}{C}}-(CH_2)_{16}-CH_3 \\ | \\ CH-O-\overset{O}{\overset{\|}{C}}-(CH_2)_{16}-CH_3 \ + \ 3\ NaOH \ \longrightarrow \ \begin{array}{c} H \\ | \\ H-C-OH \\ | \\ H-C-OH \\ | \\ H-C-OH \\ | \\ H \end{array} \ + \ 3\ CH_3(CH_2)_{16}COONa \\ | \\ CH_2-O-\overset{O}{\overset{\|}{C}}-(CH_2)_{16}-CH_3 \end{array}$$

tristearin   glycerin   sodium stearate (a soap)

Salts of the long-chain fatty acids are called soaps. This type of reaction is called saponification, particularly as it is applied to the hydrolysis of the neutral lipids.

Glycerin, a clear oily liquid with a sweet taste, has a number of important uses. It is miscible with water in all proportions and not very soluble in organic solvents. It is used extensively in the cosmetic industry and also for drug preparations.

Glycerin, either in the free or esterified form, as occurs in the triglycerides, can be dehydrated:

54

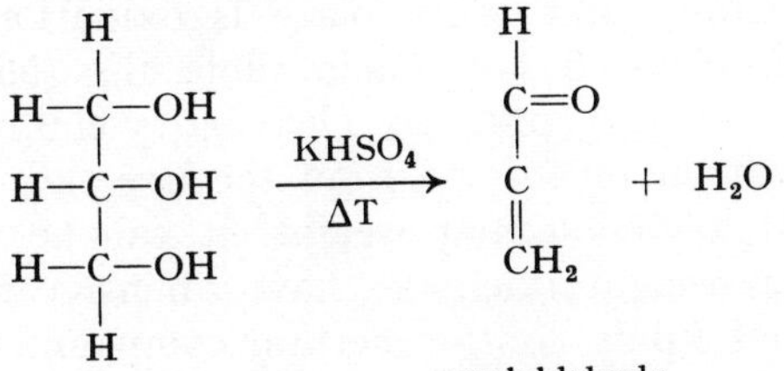

$$\begin{array}{c} H \\ | \\ H-C-OH \\ | \\ H-C-OH \\ | \\ H-C-OH \\ | \\ H \end{array} \xrightarrow[\Delta T]{KHSO_4} \begin{array}{c} H \\ | \\ C=O \\ | \\ C \\ \| \\ CH_2 \end{array} + H_2O$$

acrylaldehyde (acrolein)

The product of the reaction, acrylaldehyde or acrolein, has a very characteristic odor and is easily identified on the basis of this smell. This is a very useful test for the detection of glycerin.

Glycerin is not only a product of saponification, but it is also an intermediate in metabolism. It has the same nutritive value as carbohydrates on a weight basis.

The margarine industry converts oils, neutral lipids with low melting points, to solid or semisolid fats by hydrogenation of the unsaturated fatty acids in the molecules.

55

$$\begin{array}{l}
CH_2-O-\overset{O}{\underset{\|}{C}}-(CH_2)_7CH=CH-(CH_2)_7CH_3 \\
CH-O-\overset{O}{\underset{\|}{C}}-(CH_2)_7CH=CH-(CH_2)_7CH_3 \\
CH_2-O-\overset{O}{\underset{\|}{C}}-(CH_2)_7CH=CH-(CH_2)_7CH_3
\end{array}$$

trioelein

$$\downarrow 3H_2$$

$$\begin{array}{l}
CH_2-O-\overset{O}{\underset{\|}{C}}-(CH_2)_7CH_2-CH_2-(CH_2)_7CH_3 \\
CH-O-\overset{O}{\underset{\|}{C}}-(CH_2)_7CH_2-CH_2-(CH_2)_7CH_3 \\
CH_2-O-\overset{O}{\underset{\|}{C}}-(CH_2)_7CH_2-CH_2-(CH_2)_7CH_3
\end{array}$$

tristearin

After exposure to air for a period of time, particularly at elevated temperatures, most triglycerides develop unpleasant odors and flavors. This is called rancidification, and the oils or fats become rancid. The change that takes place is oxidation of the double bonds in the unsaturated fatty acid molecules (**56**). The peroxide which is produced can undergo cleavage yielding two aldehyde groups. If the aldehyde formed from the free end of the fatty acid chain (R_2) is of low molecular weight, it will be volatile. Most of the low-molecular-weight aldehydes have offensive odors and tastes. Many commercial lipids contain certain compounds called antioxidants, which prevent this type of reaction while the lipids are stored.

56

$$R_1-\underset{|}{\overset{H}{C}}=\underset{|}{\overset{H}{C}}-R_2 + O_2 \longrightarrow R_1-\underset{\underset{O}{|}}{\overset{\overset{H}{|}}{C}}-\underset{\underset{O}{|}}{\overset{\overset{H}{|}}{C}}-R_2$$

peroxide

$$\downarrow$$

$$R_1-\underset{\underset{O}{\|}}{\overset{\overset{H}{|}}{C}} + \underset{\underset{O}{\|}}{\overset{\overset{H}{|}}{C}}-R_2$$

As indicated elsewhere, the naturally occurring triglycerides generally are difficult to characterize by the usual parameters. Two techniques used to obtain average properties of the triglycerides are: (1) the iodine number, and (2) the saponification number. The iodine number represents the amount of iodine taken up by a given quantity of a triglyceride and is a measure of the unsaturation which is characteristic of the unsaturated fatty acids. A fat which has little unsaturation will take up little iodine and therefore has a low iodine number. On the other hand, fat with a high degree of unsaturation will have a large iodine number because it will be able to add a larger amount of iodine. The formula below shows that unsaturated fatty acids in triglyceride molecules react with iodine to form diiodo addition compounds.

57

$$-CH=CH- + I_2 \longrightarrow -\underset{\underset{I}{|}}{CH}-\overset{\overset{I}{|}}{CH}-$$

By determining the saponification number, a measure of the average chain length of the fatty acids can be obtained. The saponification number is defined as the milligrams of KOH required to saponify 1 g of fat. One mole of triglyceride requires three moles of sodium hydroxide for saponification (**53**). A triglyceride containing short-chain fatty acids has a low molecular weight. A given weight of that triglyceride will have more ester linkages and therefore will utilize more sodium hydroxide in saponification. On the other hand, the same weight of a triglyceride having long-chain fatty acids and a high molecular weight will have fewer ester linkages, resulting in a lower saponification number. In summary, a triglyceride with a high saponification number contains short fatty acid chains, whereas a fat with a low saponification number will have longer fatty acid molecules.

Some unsaturated fatty acids are apparently required in the diet of certain mammals. Obviously, this is because the organisms are not able to synthesize certain unsaturated fatty acids or at least are not able to synthesize enough to meet their needs. Laboratory demonstrations have shown that animals maintained on diets deficient in unsaturated fatty acids develop skin lesions. These lesions can be cured or at least reversed when the essential unsaturated fatty acids are returned to the diet. Recently, Madison Avenue has promoted unsaturated fatty acids as essential requirements for the average American diet. The term favored by the advertising industry is polyunsaturated fats or, more succinctly, "polyunsaturates." However, in humans, the actual requirements, if any, for unsaturated fatty acids have never been demonstrated conclusively.

Waxes are similar in structure and properties to the triglycerides. For example

58

$$CH_3(CH_2)_{12}CH-O-\overset{\overset{\displaystyle O}{\|}}{C}-(CH_2)_{14}CH_3$$
myricyl palmitate

Waxes are esters of the fatty acids but differ from the triglycerides in that the alcohols are monohydroxy rather than trihydroxy. Myricyl palmitate is one of the major ingredients in beeswax.

PHOSPHOLIPIDS

This class is characterized by the presence of phosphorous and nitrogen in the molecules. Many of the phospholipids are derivatives of phosphatidic acid.

59

$$\begin{array}{l} H_2C-O-\overset{\overset{\displaystyle O}{\|}}{C}-R_1 \\ |\\ CH-O-\overset{\overset{\displaystyle O}{\|}}{C}-R_2 \\ |\\ H_2C-O-\overset{\overset{\displaystyle O}{\|}}{P}-OH \\ | \\ OH \end{array}$$
phosphatidic acid

Phosphatidic acid, as such, does not usually occur free in living systems but usually as the phosphate diester with an amino alcohol.

60

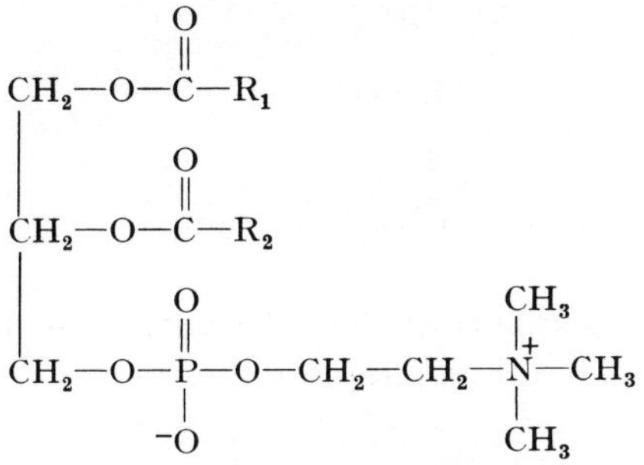

$$\begin{array}{l} CH_2-O-\overset{O}{\overset{\|}{C}}-R_1 \\ CH_2-O-\overset{O}{\overset{\|}{C}}-R_2 \\ CH_2-O-\overset{O}{\underset{-O}{\overset{\|}{P}}}-O-CH_2-CH_2-\overset{CH_3}{\underset{CH_3}{\overset{|+}{N}}}-CH_3 \end{array}$$

phosphatidyl choline
(a lecithin)

Here the amino alcohol is called choline.

61
$$HO-CH_2-CH_2-\overset{+}{N}-(CH_3)_3$$
choline

Phosphatidyl choline, like the term triglyceride, represents a large number of different compounds, since the phosphatidyl cholines which occur in nature contain a mixture of fatty acid residues. In formula **60**, R_1 and R_2 represent the fatty acid residues. Other compounds can be esterified with the phosphate groups, such as ethanol amine, and in some cases with amino acids such as serine.

62
$$HO-CH_2-CH_2-NH_2$$
ethanol amine

63 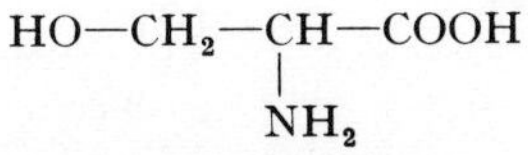
$$HO-CH_2-\underset{NH_2}{\overset{}{CH}}-COOH$$
serine

Although several other phospholipids contain glycerol, some do not. For instance, there are phospholipids which contain an eighteen-carbon aminoalcohol called sphingosine.

64 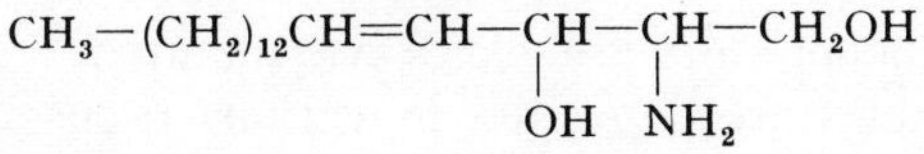
$$CH_3-(CH_2)_{12}CH=CH-\underset{OH}{\overset{}{CH}}-\underset{NH_2}{\overset{}{CH}}-CH_2OH$$
sphingosine

Phospholipids containing sphingosine are called sphingomyelins.

65

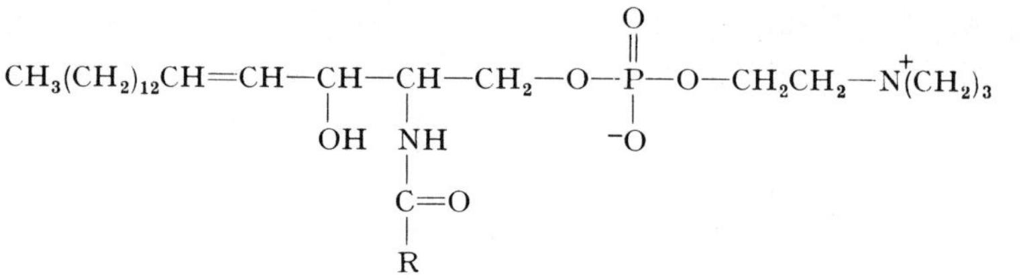

$$CH_3(CH_2)_{12}CH=CH-\underset{OH}{CH}-\underset{\underset{\underset{R}{C=O}}{NH}}{CH}-CH_2-O-\underset{\underset{-O}{\overset{\overset{O}{\|}}{P}}}{}-O-CH_2CH_2-\overset{+}{N}(CH_2)_3$$

a sphingomyelin

They differ from phosphatidic acid derivatives in that they contain only one fatty acid residue, represented as $R-\underset{\underset{O}{\|}}{C}-$ in **65,** and two aminoalcohols. Also, phospholipids can contain dihydrosphingosine. This occurs when the double bond between carbons 10 and 11 of sphingosine is reduced.

COMPOUND LIPIDS

The compound lipids comprise the third class to be considered. This group contains lipids in combination with either proteins (called lipoprotein or proteolipid) or carbohydrates (called glycolipids).

The carbohydrates in glycolipids are usually galactose or glucose. The glycoside cerebroside normally contains a predominance of the hexose β-D-galactopyranoside. As shown in **66** R represents a fatty acid.

66

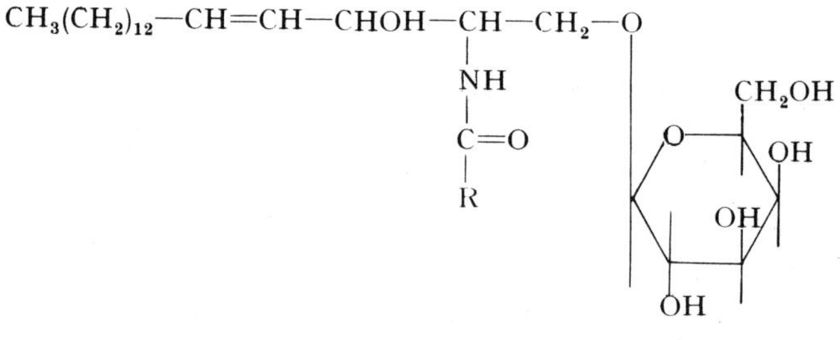

$$CH_3(CH_2)_{12}-CH=CH-CHOH-\underset{\underset{\underset{R}{C=O}}{NH}}{CH}-CH_2-O-\text{[galactopyranose ring with CH}_2\text{OH, OH, OH, OH]}$$

a cerebroside

However, in Gaucher's disease excessive quantities of glucose containing cerebrosides are deposited in various tissues.

Large amounts of glycosides also occur in the brain and in the sheaths surrounding nerve tracts. The loss of these nerve sheaths is characteristic of demyelinating diseases.

STEROLS

67 Included in the group of compounds called lipids are some diverse structures such as cyclopentanoperhydrophenanthrene.

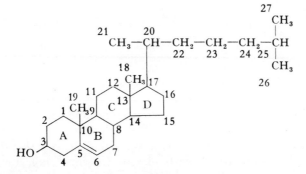

cyclopentanoperhydrophenanthrene

This is the parent compound for a large number of substances of great biologic importance. These compounds as a group are usually referred to as the sterols, a word derived from the Greek meaning solid alcohols. Some of these sterols are elaborated by the endocrine glands and constitute a group of compounds often referred to as the steroid hormones. The compounds of this type comprise the sterol class of lipids.

A well-known steroid compound is cholesterol.

68

cholesterol

The numbering system used with the sterols is shown in the above formula. Note that cholesterol contains a hydroxy group at carbon 3, and methyl groups arising from carbons 10 and 13. These methyl groups are numbered 19 and 18 respectively. Cholesterol has a side chain originating from carbon 17 with the first carbon in the chain numbered 20. The lettering system used to distinguish the four rings in the sterol structure is also shown.

Cholesterol is a lipid material found apparently in all mammalian cells. It seems to be produced and present in quantities in excess of the organisms' needs. It is known that it can serve as a precursor for steroid hormones, bile acids, and vitamin D.

69

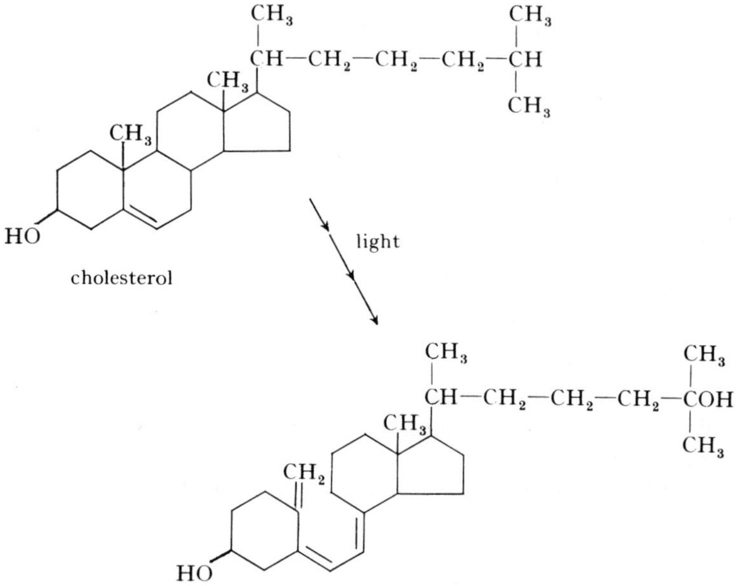

cholesterol

↓ light

active vitamin D
(25-hydroxycholecalciferol)

Cholesterol normally occurs in serum at concentrations between 100 and 280 mg/dl. In recent years considerable emphasis has been placed on the relationship between cholesterol and atherosclerosis and other forms of vascular disease. The relationship between cholesterol and heart or vessel disease is not at all clear, but there is some suggestion that individuals with elevated serum cholesterol levels tend to be more prone to these diseases than individuals with lower cholesterol levels.

CHAPTER 3

AMINO ACIDS AND PROTEINS

Proteins are undoubtedly the most complex biologic molecules and also probably the most important structures of biologic systems. Proteins are constituents of all cells. They serve as structural components, enzymes, food, hormones, antibodies, and fill many other vital roles. Before considering the proteins as such, the amino acid monomeric units will be considered.

AMINO ACIDS

As the name implies amino acids are acids carrying amino groups. In biologic systems, most of the amino acids encountered are carboxylic acids. The general formula for the amino acids is

70
$$R-\underset{NH_2}{\overset{H}{\underset{|}{C}}}-COOH$$

The degradation of proteins to the constituent amino acid monomers yields some 20 different amino acids. Glycine is the simplest amino acid.

71
$$NH_2-\underset{H}{\overset{H}{\underset{|}{C}}}-COOH$$

α-aminoacetic acid
glycine

35

Another monocarboxylic aliphatic amino acid is alanine.

72
$$CH_3-CH-COOH$$
$$|$$
$$NH_2$$

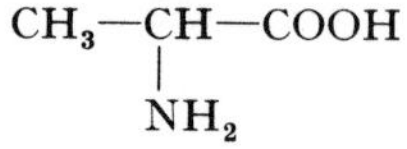

α-aminopropionic acid
alanine

There are also branched-chain aliphatic amino acids, such as valine.

73
$$CH_3$$
$$|$$
$$CH-CH-COOH$$
$$||$$
$$CH_3NH_2$$

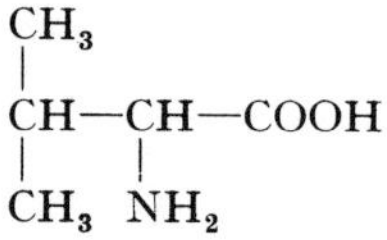

α-aminoisovaleric acid
valine

Serine represents an aliphatic amino acid having a hydroxy group. Recall that serine has been encoutered as a constituent of the phospholipids (**63**).

74
$$HO-CH_2-CH-COOH$$
$$|$$
$$NH_2$$

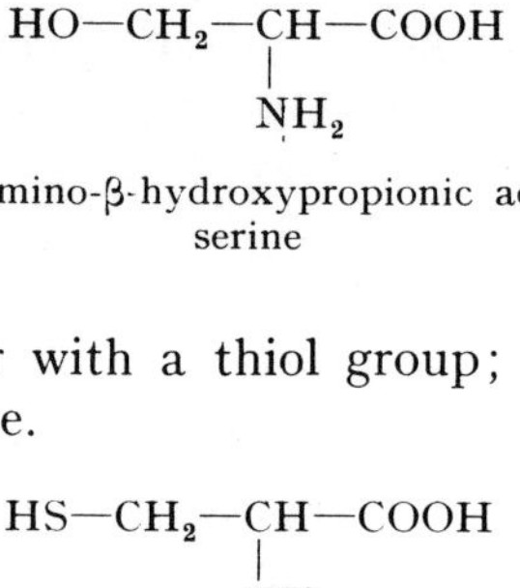

α-amino-β-hydroxypropionic acid
serine

Amino acids also occur with a thiol group; the sulfur analogue of serine is called cysteine.

75
$$HS-CH_2-CH-COOH$$
$$|$$
$$NH_2$$

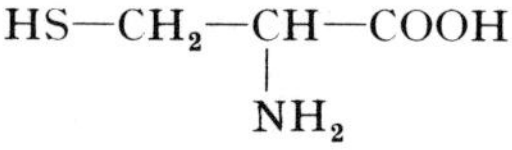

cysteine

The disulfide amino acid, cystine, is formed by the oxidation of cysteine.

76

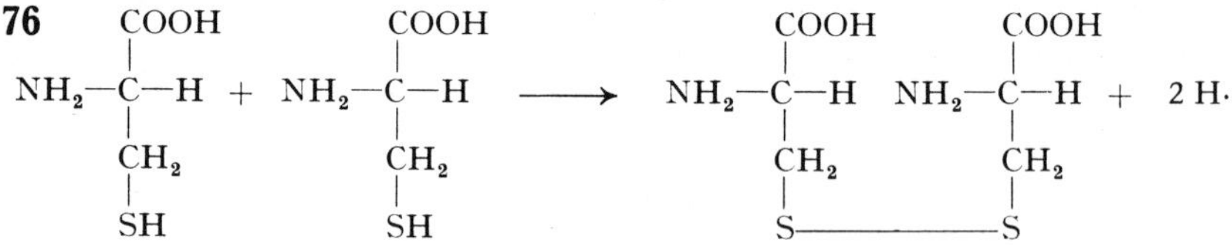

$$\begin{array}{cccc}
COOH & COOH & COOH & COOH \\
| & | & | & | \\
NH_2-C-H\ +\ NH_2-C-H & \longrightarrow & NH_2-C-H\quad NH_2-C-H\ +\ 2\,H\cdot \\
| & | & | & | \\
CH_2 & CH_2 & CH_2 & CH_2 \\
| & | & | & | \\
SH & SH & S\text{------}S
\end{array}$$

cystine

Two amino acids are dicarboxylic: aspartic and glutamic acids.

77

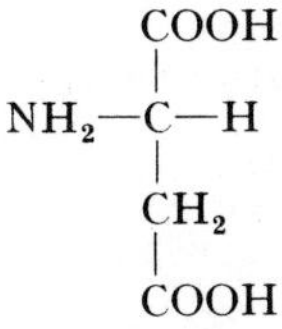

$$\begin{array}{cc} \text{COOH} & \text{COOH} \\ | & | \\ \text{NH}_2\text{—C—H} & \text{NH}_2\text{—C—H} \\ | & | \\ \text{CH}_2 & \text{CH}_2 \\ | & | \\ \text{COOH} & \text{CH}_2 \\ & | \\ & \text{COOH} \end{array}$$

aspartic acid 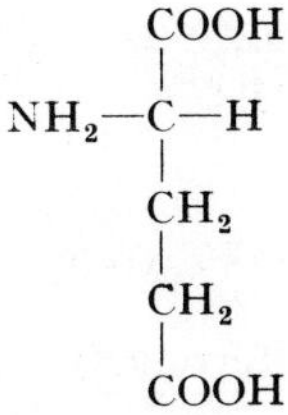   glutamic acid

As if to counterbalance this, there are diamino acids such as lysine.

78

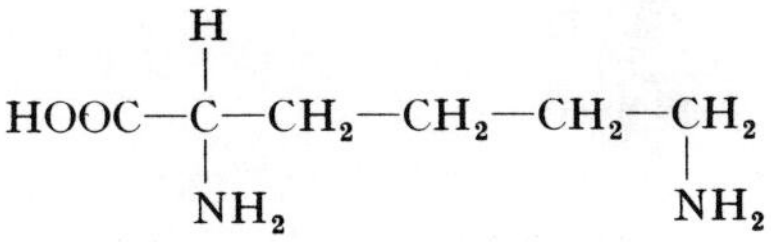

$$\text{HOOC—}\underset{\underset{\text{NH}_2}{|}}{\overset{\overset{\text{H}}{|}}{\text{C}}}\text{—CH}_2\text{—CH}_2\text{—CH}_2\text{—}\underset{\underset{\text{NH}_2}{|}}{\text{CH}_2}$$

lysine

One of the amino acids, arginine, is extremely basic because it contains a guanidino group.

79

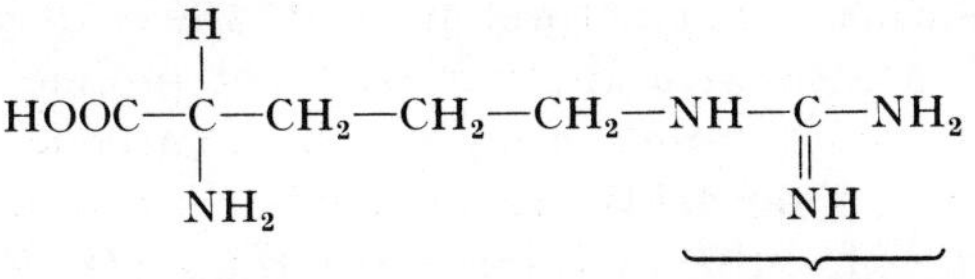

$$\text{HOOC—}\underset{\underset{\text{NH}_2}{|}}{\overset{\overset{\text{H}}{|}}{\text{C}}}\text{—CH}_2\text{—CH}_2\text{—CH}_2\text{—NH—}\underset{\underset{\text{NH}}{||}}{\text{C}}\text{—NH}_2$$

arginine   guanidino group

Attention is called to the fact that not all compounds generally considered as amino acids are indeed amino acids; one such is proline, an imino acid.

80

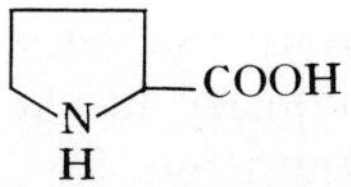

(pyrrolidine ring)—COOH
N
H
proline

Phenylalanine and tyrosine represent two of the aromatic amino acids.

81

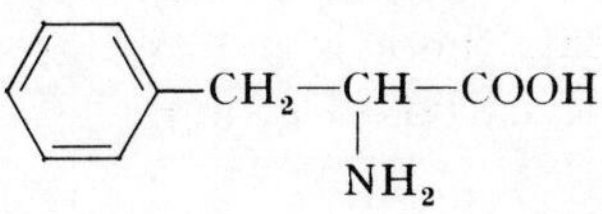

$$\text{C}_6\text{H}_5\text{—CH}_2\text{—}\underset{\underset{\text{NH}_2}{|}}{\text{CH}}\text{—COOH}$$

phenylalanine

82

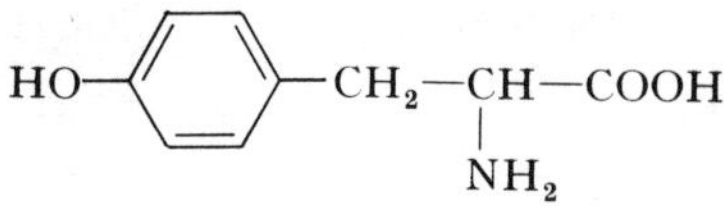

$$HO-\langle\bigcirc\rangle-CH_2-CH(NH_2)-COOH$$

tyrosine

There are also heterocyclic amino acids such as tryptophan and histidine.

83

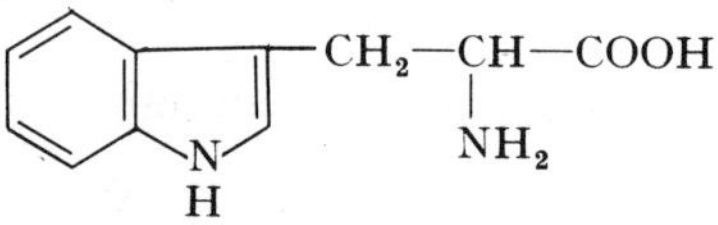

tryptophan (indole—$CH_2-CH(NH_2)-COOH$)

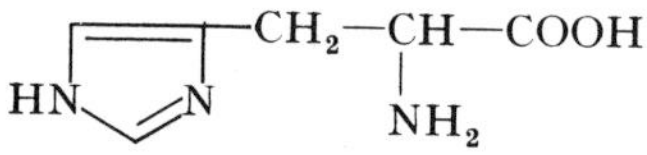

84

histidine (imidazole—$CH_2-CH(NH_2)-COOH$)

After reviewing the structures presented here, it will be obvious that all amino acids – with the exception of proline, which has an imino group – have an amino group on the α carbon. As a generalization then, the amino acids are referred to as alpha amino acids. Note also that with the first four structures (**71, 72, 73, 74**) the systemic name was given in addition to the common name. The convenience of learning the common rather than the systemic names was probably clear as more amino acids were presented. Most workers in biochemistry describe the amino acids by their common names.

Further examination of these structures reveals that each one, with the exception of the first, glycine, contains asymmetric carbon atoms. Fortunately, the majority of the naturally occurring amino acids are all of the L configuration.

REACTIONS

There are a number of interesting reactions involving amino acids. One is the reaction with acids and bases. Within the amino acid molecule shown (**85**), there is a basic group, an amino group, and an acidic group, the carboxyl group.

85

$$NH_2-CH(R)-COOH \rightleftharpoons NH_3^+-CH(R)-COO^-$$

The intramolecular reaction occurs in solution so that amino acids always exist in an ionic form. Therefore, the previous structures for amino acids (**70** through **86**) are actually incorrect. They should be shown as zwitterions or as specific salts.

For instance, if an amino acid is isolated from a HCl solution it would be a hydrochloride salt.

86
$$Cl^- \; NH_3^+\!-\!\underset{\underset{R}{|}}{CH}\!-\!COOH$$

amino acid hydrochloride

Similarly, if it were isolated from a NaOH solution it would be a sodium salt.

87
$$NH_2\!-\!\underset{\underset{R}{|}}{CH}\!-\!COO^- \; Na^+$$

In general, amino acids are not shown in the correct ionic or salt form. This improper representation is due to the fact that it is more convenient to represent the structures as nonionized molecules. This is particularly true in the writing of reactions, since accounting for the various ionic species is thereby eliminated. This type of incorrect representation is common among biochemists. Compounds such as amino acids that form intramolecular salts are called zwitterions, which from the German means "bastard ion." When a zwitterion is treated with a stronger acid the carboxyl group will accept a proton, neutralizing its negative charge, and a net positively charged amino acid results.

88
$$NH_3^+\!-\!\underset{\underset{R}{|}}{CH}\!-\!COO^- + H^+ \longrightarrow NH_3^+\!-\!\underset{\underset{R}{|}}{CH}\!-\!COOH$$

Similarly, reaction of the zwitterion with a base will result in the removal of the amino group proton, forming a negatively charged ion.

89
$$NH_3^+\!-\!\underset{\underset{R}{|}}{CH}\!-\!COO^- + OH^- \longrightarrow NH_2\!-\!\underset{\underset{R}{|}}{CH}\!-\!COO^- + H_2O$$

Thus, the form and molecular charge on the amino acid salt is regulated by the hydrogen ion concentration or pH of the solution.

The carboxyl groups of amino acids can be esterified.

90
$$NH_2\!-\!\underset{\underset{R}{|}}{CH}\!-\!COOH + EtOH \longrightarrow NH_2\!-\!\underset{\underset{R}{|}}{CH}\!-\!COOEt + H_2O$$

Furthermore, the amino group can be acylated, here with acetyl chloride, to yield the N-acetyl derivative.

91
$$\text{AcCl} + \text{NH}_2-\underset{\underset{R}{|}}{\text{CH}}-\text{COOH} \longrightarrow \text{Ac}-\text{NH}-\underset{\underset{R}{|}}{\text{CH}}-\text{COOH} + \text{HCl}$$

In general, then, the amino and carboxyl groups of amino acids can form derivatives the same as simple carboxylic acids and amines.

Amino acids react with nitrous acid to yield the unstable diazonium compound, which then undergoes spontaneous decomposition to give the corresponding alcohol.

92
$$\text{NH}_2-\underset{\underset{R}{|}}{\text{CH}}-\text{COOH} + \text{HONO} \longrightarrow \left[\overset{+}{\text{N}}=\text{N}-\underset{\underset{R}{|}}{\text{CH}}-\text{COOH}\right] \longrightarrow$$

$$\text{N}_2 + \text{HO}-\underset{\underset{R}{|}}{\text{CH}}-\text{COOH} + \text{H}_2\text{O}$$

This reaction is of considerable importance in biochemistry because it can be used to quantitate amino acids. The amount of amino acid can be determined by measuring how much nitrogen is released, as there is one mole of nitrogen released for each mole of amino nitrogen. It is a nonspecific technique and measures the total number of amino groups present and does not distinguish one amino group from another.

As indicated, amino acids cannot be quantitated by titration as acids and amines in the usual manner because of zwitterion formation. The Sörensen formal titration technique can be used, however, to obtain accurate titration values.

93
$$\text{NH}_2-\underset{\underset{R}{|}}{\text{CH}}-\text{COOH} + \text{HCHO} \longrightarrow \text{HOCH}_2-\text{NH}-\underset{\underset{R}{|}}{\text{CH}}-\text{COOH}$$

N-methylol amino acid

$$\downarrow \text{HCHO}$$

$$\underset{\text{HOCH}_2}{\overset{\text{HOCH}_2}{>}}\!\!\text{N}-\underset{\underset{R}{|}}{\text{CH}}-\text{COOH}$$

N,N-dimethylol amino acid

When two moles of formaldehyde react with an amino acid, it forms a dimethylol derivative which makes the nitrogen completely unreactive to acids and bases leaving the carboxy group free to be titrated like any other carboxylic acid. This type of reaction is often used to measure the appearance of amino acids when they are released in solution by enzymatic hydrolysis of proteins.

A very useful reaction of amino acids is that with ninhydrin. The initial reaction causes an oxidative decarboxylation of the amino acid to yield the corresponding aldehyde, carbon dioxide (CO_2), and ammonia. The ninhydrin is reduced to hydrindantin. This first reaction can be used to quantitate amino acids by measuring the CO_2 produced by the reaction, since there is one mole of CO_2 produced for each carboxyl group of the amino acid. Usually, however, the reaction is allowed to proceed (**94**). Hydrindantin, ammonia, and another mole of ninhydrin react to form a condensation product. In an alkaline medium the product forms a salt which has an intense purple color. The color can be measured colorimetrically for quantitation or detected visually for qualitative purposes.

This is a very sensitive reaction and is used extensively in amino acid chromatography.

There are many other reactions of amino acids: some are general reactions and some are specific for certain amino acids. Most of these are beyond the scope of this chapter.

94

$$NH_2\text{—}\underset{\underset{R}{|}}{CH}\text{—}COOH \;+\; \underset{\text{ninhydrin}}{\text{[structure]}} \;\longrightarrow$$

$$R\text{—}\overset{O}{\overset{\|}{C}}H \;+\; CO_2 \;+\; NH_3 \;+\; \underset{\text{hydrindantin}}{\text{[structure]}}$$

$$\underset{\text{hydrindantin}}{\text{[structure]}} \;+\; NH_3 \;+\; \underset{\text{ninhydrin}}{\text{[structure]}} \;\xrightarrow{-3H_2O}$$

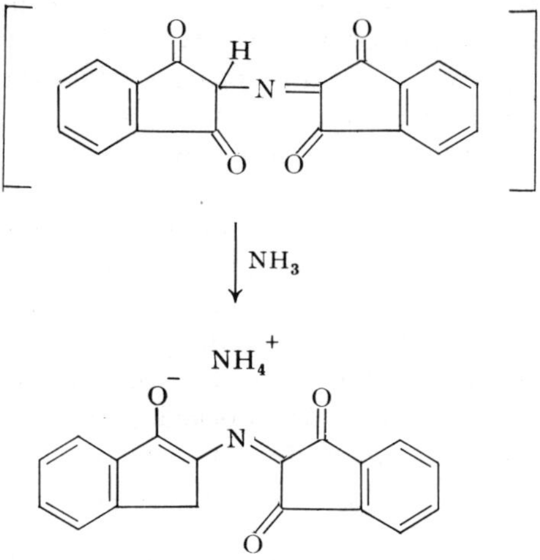

[structure with NH bridge between two indanedione units]

↓ NH₃

[structure with NH₄⁺ and purple Ruhemann's complex]

(purple)

PEPTIDES AND PROTEINS

As indicated previously, the amino acids are the monomeric units of protein — that is, proteins are polymers of amino acids. The amino acids are linked together by an amide link to form the protein polymers.

95
$$NH_2-CH(R)-C(=O)-NH-CH(R)-COOH$$
peptide

These amide linkages between two or more amino acids are called peptide linkages or peptide bonds. Peptide linkage between the two molecules of glycine result in a compound called glycylglycine.

96
$$NH_2-CH_2-C(=O)-NH-CH_2-COOH$$
glycylglycine

As shown below, three amino acids connected together by peptide links produce glycylalanylphenylalanine.
Obviously, the structures become complex when only a few amino acids are involved.

97

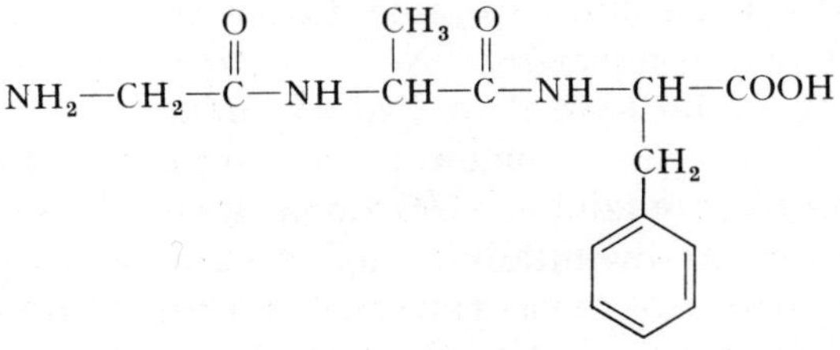

$$NH_2-CH_2-\overset{\overset{O}{\|}}{C}-NH-\overset{\overset{CH_3}{|}}{CH}-\overset{\overset{O}{\|}}{C}-NH-\underset{\underset{C_6H_5}{CH_2}}{CH}-COOH$$

glycylalanylphenylalanine

The molecular weights of typical proteins range between 6,000 and 400,000. The average molecular weight of the 20 naturally occurring amino acids is about 110. By dividing the protein molecular weights by the average amino acid molecular weight, it is found that there are 54 to 3,636 amino acids per protein molecule. In addition, most proteins are not composed of all 20 amino acids in random numbers. For instance, silk fibroin contains mostly glycine, alanine, and serine, and salmine consists mostly of arginine.

So far proteins have been considered in terms of the number of amino acids in the molecule. Of more importance is the sequence of the amino acids (refer to **97**). Note that the molecule contains three amino acids linked by peptide bonds. The sequence of the three amino acids can vary.

98

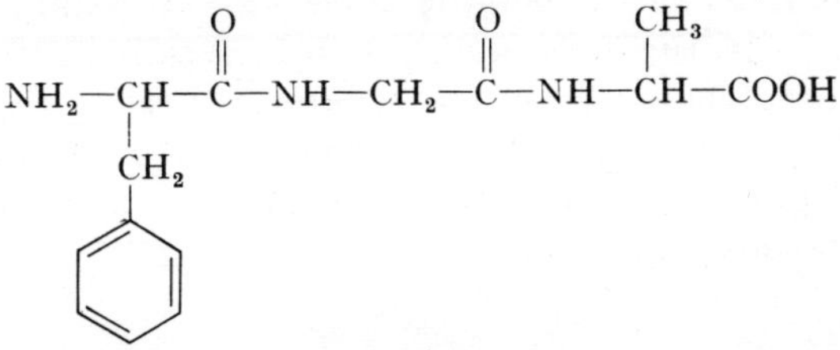

$$NH_2-\underset{\underset{C_6H_5}{CH_2}}{CH}-\overset{\overset{O}{\|}}{C}-NH-CH_2-\overset{\overset{O}{\|}}{C}-NH-\overset{\overset{CH_3}{|}}{CH}-COOH$$

phenylalanylglycylalanine

Perhaps the importance of the amino acid sequence can be better understood if the possible arrangements of three amino acids are outlined as follows:

99

A—B—C
B—C—A
C—A—B
B—A—C
C—B—A
A—C—B

Each arrangement yields a tripeptide; thus with three amino acids we have six different compounds. Now consider how many different possibilities we would have if we had 54 amino acids for one of the smaller, simpler proteins, or indeed the number of possible compounds if we had a protein containing 3,600 amino acids. It is readily appreciated that there are essentially an infinite number of possible combinations of amino acids to yield an equal number of protein molecules.

The nomenclature used to describe polypeptides can often be confusing. In **96,** glycylglycine is known as a dipeptide, but in reality it has only one peptide linkage. Similarly in **97** and **98** the two compounds are referred to as tripeptides, but there are only two peptide bonds. The prefix – tri, di, or whatever – is not determined by the number of peptide bonds but rather by the number of amino acids present in the molecule.

The di- and tripeptides (**96, 97,** and **98**) have a carboxyl group on one end and an amino group on the other very much like the simple amino acids (**71** and **72**). As might be expected, the amino acids in a polypeptide (that is, those with free carboxyl and amino groups) behave as amino acids in that they demonstrate zwitterion formation.

100 $\qquad NH_2-Pr-COOH \rightleftharpoons NH_3^+-Pr-COO^-$

The Pr, separating the amino group and the carboxyl group, represents the remainder of the peptide molecule. In reality, proteins are not as simple as pictured, because it will be recalled that some amino acids have more than one amino group (such as lysine), and some have more than one carboxyl group (such as aspartic acid). This is depicted in Figure 3-1, where the polypeptide chain shows additional amino and carboxyl groups.

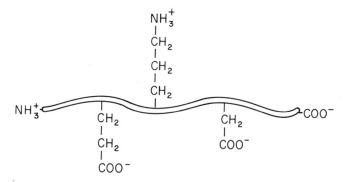

Fig. 3-1. Polypeptide showing amino and carboxyl groups along the chain in addition to the terminal groups.

When the amino and carboxyl groups are considered in addition to the N and C terminals, the protein and polypeptide can be represented as

101 $\qquad (NH_3^+)_x-Pr-(COO^-)_y$

The terms N and C terminals refer to the ends of the molecule which bear a free amino and a carboxyl group, respectively. In formulas **97** and **98,** the N terminal is on the left and C terminal is drawn on the right.

A protein may have the same number of amino groups as carboxyl groups, in which case x = y, or they may be different. A protein may have more free amino groups than carboxyl groups; then it is spoken of as a basic protein, and when dissolved in water yields an alkaline pH. By contrast, other proteins contain more carboxyl groups than amino groups; these are called acidic proteins and demonstrate an acid reaction when dissolved in water. The zwitterion form of protein behaves similarly to that of the amino acids. In an acid medium the polypeptide reacts to give a positively charged molecule.

102 $\quad (NH_3^+)_x-Pr-(COO^-)_y + yH^+ \longrightarrow (NH_3^+)_x-Pr-(COOH)_y$

Reaction of the protein zwitterion with alkali leads to a negatively charged molecule.

103 $\quad (NH_3^+)_x-Pr-(COO^-)_y + xOH^- \longrightarrow (NH_2)_x-R-(COO^-)_y$

A polypeptide in a strongly acid medium is fully protonated and therefore is a positively charged molecule (**102**). Conversely, in a strongly alkaline medium a polypeptide will be a negatively charged molecule (**103**). In an acid solution, the number of charges per molecule is equal to the number of amino groups, or x, and in an alkaline medium the number of molecular charges is equal to the number of carboxyl groups, or y.

In a solution where the proton concentration is limited (that is, $H^+ < y$), only part of the carboxyl groups would be protonated. However, the amino groups would remain protonated. It is apparent that the charge on the protein molecule is regulated by the hydrogen ion concentration or the pH of the solution. Further, it can be realized that by proper regulation of the pH, a hydrogen ion concentration can be obtained where the number of negative charges and the number of positive charges are equal, so the overall net charge is zero – that is, the sum of the negative and positive charges are equal and opposite, thus the net charge on the molecule is zero. The pH at which the number of negative charges and the number of positive charges are equal is known as the isoelectric point or the isoelectric pH. It should be added that the term isoelectric point is not restricted to proteins and polypeptides but also applies to amino acids.

ELECTROPHORESIS

The fact that the charge on a protein can be regulated is often useful in that it allows the separation of certain proteins from each other and from other types of compounds. A device for separating proteins according to their electrical charge has electrodes immersed in a conducting medium so that an electrical current can pass through the solution as a result of a potential difference across the electrodes. This type of separation is referred to as electrophoretic (see Fig. 3-2).

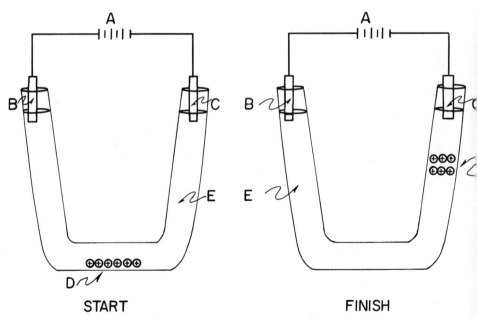

Fig. 3-2. Schematic representation of an electrophoresis system. A, battery ; B, anode; C, cathode; D, protein molecules; E, conducting medium.

In the electrophoresis apparatus, positively charged protein molecules are introduced at the starting point and a potential difference is applied across the electrodes. The positively charged protein molecules will migrate into the area of the cathode. Thus, electrophoresis can be used to separate positively charged molecules from other molecules. However, it is not ordinarily used for this purpose.

Usually electrophoresis is used to characterize proteins by their rate of migration or mobility. The rate at which a protein will migrate in an electric field depends on several factors: the total net number of charges on the molecule, the size, and to some extent, the shape of the molecule.

The type of electrophoresis illustrated, called free boundary, is rarely used today because it requires elaborate equipment and considerable space. Electrophoretic separations can be accomplished in many types of media besides solutions. Filter paper and cellulose acetate saturated with conducting media have found extensive application. Various gels on solid supports have been used, including starch, agar-agar, and polyacrylamide. Paper electrophoresis is schematically represented in Figure 3-3.

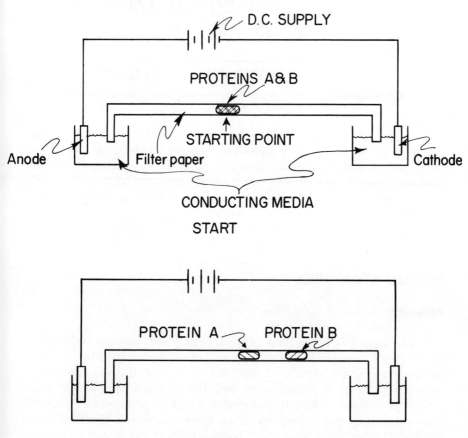

Fig. 3-3. Filter paper electrophoresis.

Electrophoretic techniques are used extensively for serum protein separations. Although serum contains a large number of different proteins, the proteins can usually be separated into five distinct groups on the basis of their mobility in an electric field at pH 8.6. Represented schematically, albumin migrates the fastest, and thus the furthest, during a period of separation (see Fig. 3-4).

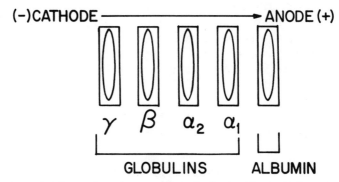

Fig. 3-4. Electrophoretic separation obtained with normal serum showing the five components.

The α_1-, α_2-, β-, and γ-globulins respectively, follow in terms of decreasing mobility. The γ-globulin fraction, depending upon the particular situation, either does not migrate or perhaps actually migrates toward the cathode. After the proteins are separated according to their mobility, the quantity of each fraction can be measured in various ways. Electrophoresis provides a means for separating and quantitating the five fractions of serum proteins.

PROTEIN STRUCTURES

PRIMARY

The structure of proteins is apparently much more complex than might be expected. The primary structure is determined by covalent intrapeptide chain bonds. These bonds include the peptide bonds between individual amino acids and the disulfide bridges between cysteinyl residues. Recall that cysteine can be oxidized to form cystine, which is two molecules of cysteine connected together by a disulfide bridge **(76)**. This same event can take place when cysteine is present in protein molecules forming a disulfide bridge within a single or between two peptide chains.

104

```
        C=O                  C=O
        |                    |
   H—C—CH₂—S—S—CH₂—CH
        |                    |
        NH                   NH
        |                    |
        C=O                  C=O
```

One type of configuration conferred by intrapeptide chain disulfide bonds can be represented schematically (see Fig. 3-5).

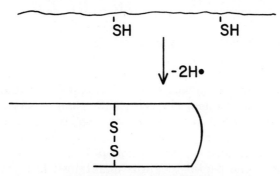

Fig. 3-5. Configuration that results from formation of an intrapeptide disulfide bond.

Initially, we have a peptide chain with two sulfhydryl groups arising from cysteine groups; but upon oxidation the two sulfhydryl groups are joined, forming a disulfide bridge, and the protein molecule is

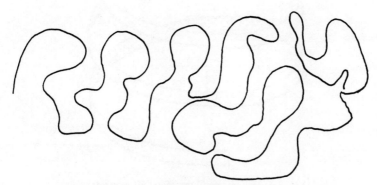

Fig. 3-6. Random coil configuration of a protein molecule.

then bent into a fishhook configuration. Some protein molecules contain several disulfide bridges and can assume various shapes.

The covalent bonds of the primary structure are flexible and free to rotate. If proteins had only primary structural bonding, the molecules would assume random coil configuration. A schematic representation of this is seen in Figure 3-6.

SECONDARY

Usually, proteins in biologic systems do not have random configurations, instead they possess very ordered and sometimes complex structures. The major part of this ordered structure is due to hydrogen bonding between carbonyl oxygen and amide hydrogen called the secondary structure.

You may recall from inorganic chemistry that some atoms have a greater attraction for electrons than others. When these atoms are lined up according to their relative affinities, we have what is often called the electromotive series. Oxygen is a rather aggressive molecule and likes to associate with more than its share of protons.

Since the peptide bonds recur at regular intervals, hydrogen bonding of the secondary structure can also occur with regularity. This regularity frequently allows the protein to assume a helical configuration. The peptide chain is maintained in a coil shape by the hydrogen bonding (see Fig. 3-7).

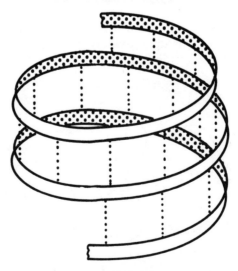

Fig. 3-7. Helical configuration of a protein molecule due to hydrogen bonding.

TERTIARY

All other types of interaction leading to additional folding or intrapeptide bonding is represented in the tertiary structure of a protein. This structure depends in part upon hydrogen bonding other than that of the amide (peptide) groups.

Other interactions between amino acid residues in peptide chains also contribute to tertiary fine structure. One type of interaction is called ionic bonding.

105.
```
         A                                              B
         .                                              .
         .                                              .
         NH                                             NH
         |                                              |
         C=O                                            C=O
         |                                              |
     H—C—CH₂—CH₂—CH₂—CH₂—NH₃⁺ . . . . OOC⁻—CH₂—CH
         |                                              |
         NH                                             NH
         |                                              |
         C=O                                            C=O
         .                                              .
         .                                              .

       lysyl                                         aspartyl
         A                                              B
```

Peptide chain A contains a positively charged lysyl residue and peptide B contains a negatively charged aspartyl group. There is the expected attraction between these two oppositely charged groups.

QUATERNARY

The term quaternary structure describes all types of bonds and interaction between different peptide chains. A schematic example is seen for bovine insulin in Figure 3-8.

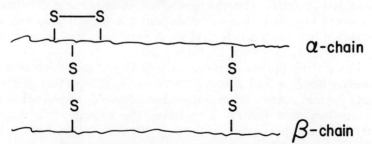

Fig. 3-8. Representation of bovine insulin showing intrapeptide disulfide bond (primary structure) and interpeptide disulfide bonds of quaternary structure.

There is an intrapeptide chain disulfide bridge (primary structure for α chain) and two interpeptide chain disulfide bridges connecting α and β chains, which are part of the quaternary structure of the molecule.

DENATURATION

Proteins can undergo profound changes in physical properties without an apparent change in chemical composition. An example of this is a cooked egg. Certainly, the characteristic properties of the egg are greatly changed. If, however, part of the egg is analyzed before cooking and compared with the chemical analysis obtained after cooking they will indeed be found the same. This sort of change in a protein is referred to as denaturation. Literally, denatured means " not native " – that is, not in the form in which it existed in the biologic system. However, in very few cases, if any, is it known exactly what proteins are like when they are in living systems. There is generally no way of accurately evaluating the properties and configurations of proteins in vivo. Since the native form is usually not known, it is difficult to know what the denatured form is. In the case of fried egg versus raw egg, there are obviously some differences and there is little confusion. However, the criteria for determining whether a protein is native or denatured is sometimes very difficult to set. This is due in part to the fact that some proteins may be denatured and then renatured or returned to what we believe to be the native state; in other words, in some cases it is a reversible process.

Denaturation involves changes in secondary, tertiary, and perhaps quaternary structures. As previously indicated, the usual type of chemical analysis will not reveal any change between a native and denatured protein. No peptide bonds are hydrolyzed, no disulfide bonds are broken, and it has the same overall composition as the native protein. The changes which occur during denaturation are limited to the hydrogen bonds and the ionic bonds and other interaction between peptide chains. A very simplified representation of denaturation is shown in Figure 3-9.

The native protein here has a helical portion which is due to hydrogen bonding and a lower " wrap-around " portion due to disulfide bridges. After denaturation has occurred, there is disruption of secondary and tertiary structures, the helical configuration is gone, and the upper portion of the molecule is represented as a random coil.

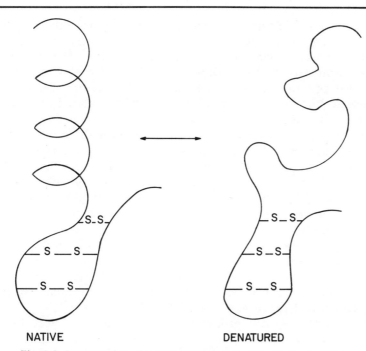

Fig. 3-9. Denaturation of a protein with loss of hydrogen bonding.

Denaturation has importance other than helping to understand what happens when you cook breakfast eggs. It is very important in biologic systems, because enzymes, which are proteins, may lose their enzymatic activity when they are denatured. Antibodies no longer function as antibodies when denatured. Hemoglobin no longer transports oxygen when denatured. There are many other examples, but these are sufficient to indicate the importance of denaturation.

These examples suggest that denaturation is a phenomenon that is quite easy to detect. In practice, this is not the case. Denaturation may be compared to poverty and beauty: poverty because there are many degrees, and beauty because it is largely in the eye of the beholder. Some proteins are very resistant to denaturation, while others are very sensitive. Instances may occur where proteins lose one form of biologic activity and retain another or indeed exhibit a different activity after denaturation.

In addition to the changes in configuration that occur with denaturation, there are obviously changes in physical properties, such as viscosity and solubility. Denatured proteins are generally less soluble than native proteins. If the treatment of proteins is severe enough (exposure to strong acids, organic solvents, heat, mechanical agitation), not only are the proteins denatured but they also coagulate and form a very insoluble precipitate. The coagulation process ordinarily requires water. Evidence for this is the fact that dried egg albumin when heated to 100° C is not denatured or coagulated, but in solution it is very easily coagulated at temperatures less than 100° C.

Intermediate between denaturation and coagulation is a process called flocculation. If a denatured protein is in a solution at its isoelectric point it usually will precipitate. All proteins are least soluble at their isoelectric points, and denatured proteins are even less soluble. Flocculation, then, is the precipitation of denatured protein at or near the isoelectric point, without coagulation. Flocculation usually occurs under mild conditions, whereas coagulation occurs under more severe conditions. Flocculation also differs from coagulation in that flocculated protein can often be redissolved by adjusting the pH away from the isoelectric point. However, a coagulated protein, once precipitated, is generally insoluble except in very strong acid or base.

As indicated earlier, proteins are of vast importance to biologic systems. They contribute not only to the structure of the organism, but they are also major components of many systems, such as enzymes, antibodies, hemoglobin, and hormones. Also, proteins or amino acids are important as a source of food.

One interesting aspect about proteins is that they generally are species specific and in some cases specific for individual animals. For instance, human serum albumin, although similar, is a different protein than bovine serum albumin; indeed bovine serum albumin serves as an antigen in humans. There are differences between proteins within species. Evidence for this is host rejection of heterotransplants.

Some types of animals are unable to make certain amino acids, or at least are unable to make enough of certain amino acids to meet their growth and maintenance requirements. This is similar to the essential fatty acids which were presented earlier. Young healthy men (" volunteer " graduate students) are able to synthesize adequate supplies of all but about eight amino acids. Since the essential amino acids cannot be provided by the organism, they must be obtained from dietary sources. Proteins for dietary use must be selected with some care, because some proteins are deficient in certain amino acids and adequate amounts of these amino acids will not be supplied regardless of the amount consumed. An example of this is a protein from corn (zein) that is deficient in tryptophan.

Animals cannot survive for long using zein as a sole source of amino acids. The diet has to be supplemented with other proteins which provide adequate amounts of tryptophan. In most parts of the world, adequate supplies of essential amino acids is a problem. This is not true in the United States, since most Americans eat a very wide variety of foods and thus are almost automatically assured of getting ample quantities of all the essential amino acids. The only individuals who are apt to have problems are those who have idiosyncrasies (food faddists) such as certain vegetarians.

CHAPTER 4

ENZYMES

Enzymes are protein catalysts. A catalyst is a substance capable of altering the rate of a chemical reaction, but which is not consumed in the reaction. It does not alter the equilibrium point, but allows equilibrium to be reached faster than it would otherwise. In the reaction shown below, A is converted to B, and in the reverse reaction B is converted to A.

106

$$A \rightleftharpoons B$$
$$\text{equilibrium}$$
$$\text{rate } A \longrightarrow B = \text{rate } B \longrightarrow A$$

At equilibrium the rate at which A is converted to B is equal to the rate at which B is converted to A. Therefore, the concentration of A and B at equilibrium will not change. Figure 4-1 indicates the

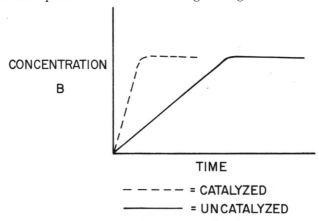

Fig. 4-1. Time-course of reaction, uncatalyzed (*solid line*) and catalyzed (*dashed line*).

difference in the increase in concentration of B with respect to time in a catalyzed and uncatalyzed reaction. Uncatalyzed, the reaction reaches equilibrium much slower than when catalyzed. Remember that enzymes do not make reactions "go," they merely allow them to reach equilibrium at different rates (usually faster) than uncatalyzed reactions.

Biologic systems are composed largely of organic compounds, and most organic reactions are slow to reach equilibrium. Elevated temperatures or pressures or both, strong acids and bases, and selective use of nonaqueous solvents are often utilized in the laboratory to compensate for this sluggishness. Since biologic systems are composed largely of organic compounds, the reactions are mostly organic. However, the conditions under which these reactions occur are extremely mild (37° C; neutral, aqueous solutions; and atmospheric pressure). Due to the presence of enzymes, it is possible for these difficult and sometimes complex reactions to take place in an environment compatible with life processes. In fact, without enzymes, the vast majority of reactions which take place would be so slow that the systems could not exist as we know them.

HISTORY

Carbohydrate fermentation to make alcohol, the souring of milk, and the production of ammonia in urine are actions of enzymes which have been known since the dawn of history. Earlier workers believed these so-called "spontaneous" reactions were associated with the life process itself. In other words, these reactions were catalyzed by living cells and could not be separated from the life processes of these cells. This concept prevailed until the early 1800s. At this time, the enzyme amylase was successfully isolated from cells and shown to have the same activity as the enzyme within living cells.

It was not until about 1900 that enzyme activity was associated with proteins. There have been several reports in the literature of enzymes which are not protein, but all of the work which has been substantiated indicates that enzyme activity is always associated with protein molecules. This does not mean that enzymes are entirely protein, since part of the molecule may be lipid, carbohydrate, or a combination of different substances. However, the active catalytic site in the molecule is apparently always protein. Some of the earlier workers felt that enzymes, though associated with proteins, had some special constituent which accounted for their catalytic properties. Some of the special constituents considered were unusual amino acids and metal ions. It is now known that the amino acids in enzymes are not different than the amino acids occurring in other proteins. There may be metals and various organic compounds associated with the molecule, but the fundamental properties reside in the polypeptide structure.

A number of enzymes have been very highly purified some to such an extent that they have been obtained in the crystalline form. Furthermore, the amino acid sequences of a number of enzymes have been determined. The results, at present, do not indicate that there is only one amino acid or that there are only certain amino acids which are responsible for enzyme activity, although serine and histidine have been associated with the active sites in several enzymes. With present knowledge, it is fair to say that all the factors involved are not known. It is known, however, that the spatial relations of the active groups is extremely important, because changes in enzyme conformation usually result in changes or losses in enzymatic activity. Recently, two independent groups of investigators assembled from amino acids the enzyme RNase (ribonuclease) and found that it possesses properties identical to the naturally occurring enzyme. Using the same techniques it will be possible to assemble structures with variations in the amino acid sequence, and from endeavors of this type most of the requirements for enzyme activity can be learned. It is also reasonable to expect in the near future that enzymes will be synthesized de novo which have no counterpart in nature. Such custom-made enzymes would have innumerable applications in industry, medicine, and biochemistry.

ENZYME ACTION

A great many studies have been made in an attempt to clarify the enzymatic mechanisms. These studies show that enzymes combine sometimes very specifically with the compound that is to undergo chemical change. This reactant compound is ordinarily termed the substrate, and the result of the reaction is usually called the product. The enzyme combines with the substrate to form what is often termed an active complex.

107 $$E + S \rightleftharpoons ES \longrightarrow E + P$$

$$E = \text{enzyme}$$
$$S = \text{substrate}$$
$$P = \text{product}$$
$$ES = \text{active complex}$$

Then the chemical change takes place in the substrate, converting it to the product. The enzyme is released to catalyze the reaction further.

COFACTORS

Some enzymes require other molecules called cofactors for activity. Cofactors can be inorganic or organic. Amylase, for instance, requires chloride ion and carbonic anhydrase must have zinc for activity. Usually, organic cofactors are called coenzymes and many are derivatives of vitamins. A coenzyme function is known for each of the water-soluble vitamins except vitamin C. Coenzymes are heat-stable and small when compared with the size of the enzyme molecules.

Most enzymes that catalyze oxidation-reduction reactions require coenzymes. An example of this is seen in the oxidation of an α-hydroxy acid to the corresponding α-keto acid.

108

$$\begin{array}{c} \text{COOH} \\ | \\ \text{H—C—OH} \\ | \\ \text{R} \end{array} + \text{NAD}^+ \rightleftharpoons \begin{array}{c} \text{COOH} \\ | \\ \text{C=O} \\ | \\ \text{R} \end{array} + \text{NAD} \cdot \text{H} + \text{H}^+$$

The coenzyme is NAD⁺, nicotinamide adenine dinucleotide. When the substrate in this reaction is oxidized the coenzyme is reduced.

In general, coenzymes are not optional enzyme accessories. Enzymes that require coenzymes are inactive without them. For activity, some enzymes require a single coenzyme whereas others have activity with similar but different coenzymes.

ENZYME MODEL

A hypothetical enzyme is schematically represented in Figure 4-2. In illustration A a small portion of the molecule is shown with the active site or catalytic center positioned adjacent to the coenzyme binding site. The position of the sites is due to the secondary and tertiary protein structures. The oxidized form of the coenzyme, in B, combines with the enzyme at its binding site. In C the substrate is shown sorbed onto the active site. At this point, D, the active enzyme catalyzes the oxidation of the substrate. This is accomplished by the loss of two protons and two electrons from the α-carbon and with the formation of a double-bonded oxygen (keto group). In E, the reduced coenzyme and end product of the

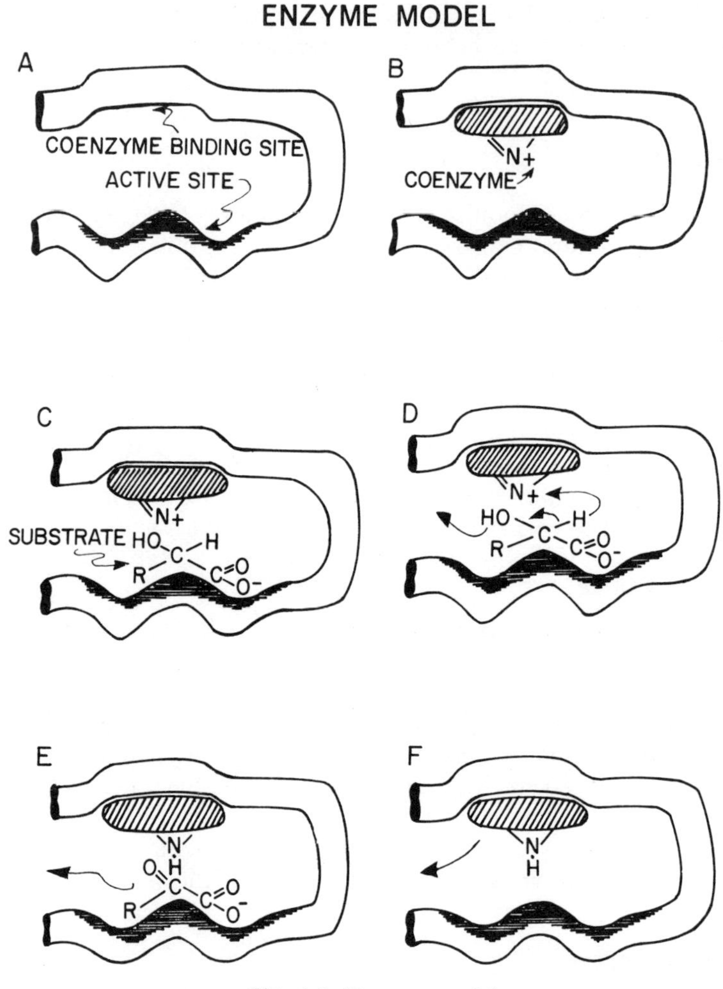

ENZYME MODEL

Fig. 4-2. Enzyme model.

reaction is seen. Since the binding affinity of the product is much less than that of the substrate it leaves the active center.

The coenzyme must be regenerated before the enzyme can catalyze the reaction of another molecule of substrate. In the case of some enzymes the binding of the coenzyme is similar to that of the substrate. That is, when the coenzyme undergoes a chemical change it is released, leaving the binding site on the enzyme available for another molecule of coenzyme in the proper chemical state. For instance, in the previous example the reduced coenzyme (NAD·H), seen in E, would be released, leaving the binding site unoccupied as in *A*. The available binding site would be occupied by an oxidized form of the coenzyme (NAD⁺) as seen in *B*.

Some enzymes, however, bind their coenzymes very tightly. In this case, regeneration of the coenzymes occurs while they are still bound to the enzymes.

Several important aspects can be illustrated by the enzyme model presented. One is the importance of the peptide chain configuration, which provides proper spatial relationship for the binding sites. As shown, there are two different binding sites (substrate and coenzyme). Apparently, however, some enzymes have a single binding site that consists of two different parts of the same peptide chain.

SPECIFICITY

Another aspect considered is the "fit" of the binding sites for the substrate and coenzyme. To some extent enzyme specificity is determined by how well the reactants fit onto the enzyme surface. Some enzymes are very specific and show activity with only one substrate. However, other enzymes are much less particular and will catalyze reactions with several similar compounds.

If the enzyme, as illustrated, were specific, it might catalyze the oxidation of lactic acid (CH_3—CH—COOH) but not α-hydroxybutyric
$\qquad\qquad\qquad\qquad\qquad\qquad\quad$|
$\qquad\qquad\qquad\qquad\qquad\qquad\:\,$OH

acid (CH_3—CH_2—CH—COOH). An enzyme with less specificity might
$\qquad\qquad\qquad\quad$|
$\qquad\qquad\qquad\;\,$OH

catalyze the reaction of both compounds as well as other members of the homologous series. Indeed, it might catalyze the reaction for a number of different compounds containing only the carboxyl and α-hydroxy groups in common.

Urease is an actual example of a very specific enzyme. It catalyzes the hydrolysis of urea.

109
$$NH_2-\overset{\overset{\displaystyle O}{\|}}{C}-NH_2 + H_2O \longrightarrow (NH_4)_2CO_3$$

Alteration in the structure of urea results in the loss of activity. For example, N-methylurea, N-nitrourea, and thiourea are not substrates.

110

$$NH_2-\overset{\overset{\displaystyle O}{\|}}{C}-NH-CH_3 \qquad NH_2-\overset{\overset{\displaystyle O}{\|}}{C}-NH-NO_2 \qquad NH_2-\overset{\overset{\displaystyle S}{\|}}{C}-NH_2$$
$\qquad$N-methylurea $\qquad\qquad\qquad$ N-nitrourea $\qquad\qquad\qquad$ thiourea

Conversely, papain, a protease, has little specificity. It will hydrolyze peptide bonds with little or no regard for the amino acid side chains.

For any chemical reaction to occur, the reactants must come together. This important aspect of enzyme action is also indicated in the model shown in Figure 4-2. The enzyme-binding sites bring the reactants together with proper alignment and spatial relationship so that the reaction is promoted.

Why are enzymes such large molecules when the binding sites generally consist of only a few amino acids? This is a justifiable question, but present knowledge does not allow an answer. It is reasonable to assume, however, that much of the molecule is concerned with maintaining the proper configuration.

COMMON ENZYME PROPERTIES

TEMPERATURE OPTIMA

Each enzyme has characteristic properties, but most demonstrate properties common to other enzymes. One of these is the effect of temperature. If the rate of an enzyme reaction is measured at various temperatures and a curve is plotted, it is generally found that the activity increases with temperature until a certain temperature is reached, generally 40 to 50° C (see Fig. 4-3).

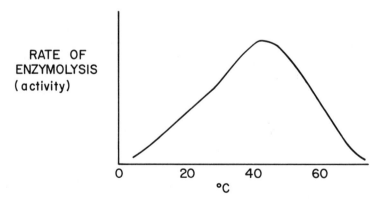

Fig. 4-3. Temperature optimum for a typical mammalian enzyme.

At higher temperatures the activity decreases. From your knowledge of chemistry you would expect the activity to increase as the temperature increases; however, it might be somewhat surprising to find that at rather mild temperatures enzyme activities actually decrease. This is due to heat denaturation of the enzyme at the higher temperatures. In Figure 4-3, at around 40° C, the rate of denaturation approaches that of the rate of enzyme reaction. At higher temperatures the rate of denaturation exceeds the rate of enzymolysis, and thus the net effect is a decreased enzymolysis.

If secondary and tertiary protein structures are recalled, it should not be difficult to visualize the effects of temperature on enzymatic activity. As the temperature of the medium increases so does the kinetic energy of the enzymes. For each enzyme a critical energy level is reached at which the configuration changes such that it becomes inactive. Small configuration changes can disrupt the spatial relationship of the active sites with the expected loss of activity.

Most enzymes from mammalian systems show a temperature optimum, 38 to 40° C, the temperature at which they show the greatest activity. This is only a generalization, however, as there are many examples of enzymes which have higher temperature optima. For example, the optimum temperature for amylase is generally around 55° C.

pH OPTIMA

The hydrogen ion concentration or pH of the solution also has an effect on the rate of enzyme reactions. Most enzymes demonstrate greatest activity at a pH of around 7 (see Fig. 4-4).

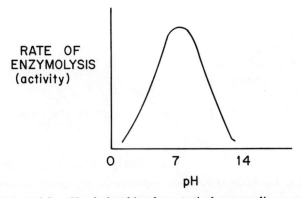

Fig. 4-4. Activity-pH relationships for a typical mammalian enzyme.

The effect of pH is due primarily to the effect that it has on the charge at the active site on the protein molecule. You recall from previous discussions that the charge on the protein molecule is directly related to the hydrogen ion concentration. Charged groups are responsible for ionic bonding of protein tertiary structure. Also, some binding sites depend upon charged groups to bind reactants. It is not surprising then that changes in hydrogen ion concentration (pH) would have a major effect on enzyme activity.

Reference to equation **107** should suggest that the charge on the enzyme molecule should have a large effect on its combination with substrate. Not only do protein molecules have charges, but also many of the substrate molecules bear charges. If the substrate was charged oppositely to that of the enzyme, the two would combine readily. However, if the charge on the substrate molecule was the same as that on the enzyme molecule, one would expect a repulsion. Also, pH has an effect on the denaturation of enzymes. Denaturation can be accomplished with strong acids and alkalis or indeed with rather mild acids and alkalis under suitable conditions. The pH optima for enzymes vary, but in general mammalian enzymes have optima around pH 7, varying somewhere between 6 and 8. However, there are notable exceptions. Examples of this would be alkaline phosphatase, which has a pH optimum of around 10, and acid phosphatase, which has a pH optimum of around 4.5.

KINETICS

Another factor which affects the rate of enzyme reactions is the concentration of substrate (see Fig. 4-5). If the rate of the reaction

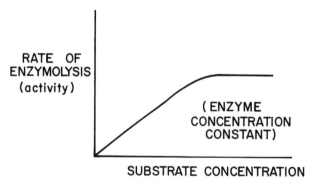

Fig. 4-5. Effect of substrate concentration on enzyme activity.

is measured at various substrate concentrations, it is found that as the substrate concentration increases, there is an increase in the rate of enzyme reaction, until a point is reached at which higher concentrations of substrate do not result in a greater rate of enzyme reaction. The explanation of this can be seen in **107.** When the rate of reaction can no longer be increased, all of the enzyme is combined with substrate. This condition is often referred to as enzyme saturation—that is, there is essentially no free enzyme to combine with the substrate.

If a slightly different experiment is done where there is an excessive concentration of substrate and the rate of enzyme reaction is measured while increasing the concentration of enzyme, a linear relationship is obtained (see Fig. 4-6). There is a linear function

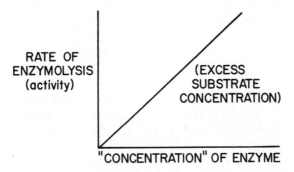

Fig. 4-6. Effect of enzyme concentration on enzyme activity.

throughout the entire range of experimental values because there is always an excess of substrate. Each point is determined by increasing the amount of enzyme. The amount of substrate present in the reaction medium is always sufficient to completely saturate the enzyme. By adding more enzyme, the rate of enzymolysis increases. This type of experiment is one of the test procedures used to determine whether a reaction is catalyzed by an enzyme. This is sometimes quite useful because there are many reactions which occur readily without enzymes, and it is difficult to determine whether the reactions are indeed catalyzed by an enzyme. If the reaction is not enzymatic, the rate of reaction would not increase with the addition of more enzyme.

ACTIVITY MEASUREMENTS

One concept that is very important in studying enzymes is the system used to indicate enzymatic reactivity or the extent of catalysis. The situation with enzymes is complex. Enzymes usually occur in very small concentrations and occur in biologic systems where there is a great deal of protein relative to the amount of enzyme protein.

Analytic techniques to specifically measure the amount of enzyme protein are not generally available. The estimation of proteins is difficult at best, and under conditions where there is a predominance of nonenzyme protein it is more difficult. Even if it were always possible to measure the concentration of the enzyme protein this would not be sufficient, as the proteins can exist in two forms: the native and the denatured. If enzyme protein could be measured, a technique for determining what portion of the total concentration was native and what portion was denatured would be required, as denatured enzymes do not have catalytic activity. To avoid the analytic problems, enzymes are usually referred to in terms of their activity or extent of catalysis rather than concentration.

The principles presented in Figures 4-5 and 4-6 are utilized in measuring enzymatic activity. When the substrate is in excess the rate of enzymolysis is constant for a given concentration of enzyme. It can be demonstrated that small quantities of enzyme produce a lower rate of reaction and larger amounts yield greater reaction rates. Therefore, the rate of enzymatic reaction is directly related to the amount of active enzyme present when the concentration of substrate is in excess. Using this relationship, enzyme activity can be determined by measuring the change in concentration of substrate or product during a fixed time period. Obviously, this is a rate measurement, since it is a change in concentration with respect to time.

An example would be the measurement of phosphatase activity in serum.

111

$$\underset{\text{substrate}}{R-O-\overset{\overset{O}{\|}}{\underset{\underset{OH}{|}}{P}}-OH} + H_2O \xrightarrow{\text{phosphatase}} R-OH + H_3PO_4$$

The protocol is to mix serum and substrate solution and incubate the mixture for a fixed period of time (30 minutes in this example) at a constant temperature and pH. Another serum aliquot is mixed with the same amount of substrate solution. In this case, the reaction is stopped immediately (zero time of reaction) by denaturing the enzyme. The concentration of inorganic phosphorus (H_3PO_4) in each sample is measured.

112   zero time   1.13 mM
          30 minutes 3.72 mM

The difference in inorganic phosphorus concentration due to release from the substrate is 2.59 mM. Thus, the reaction rate would be 2.59 mM/30 minutes.

This is an example of the so-called "two point" or steady state assay technique. Providing that the reaction rate does not change during incubation, this is a valid approach for measuring enzymatic activity. Frequently, however, the reaction rates do change, making this approach invalid. For instance, if enough substrate is consumed so that it is no longer in excess, the rate of reaction changes. Also, rate changes can occur because some enzymes are inhibited by the product produced. For example, as the reaction proceeds the product concentration increases, as does the degree of enzyme inhibition; thus the reaction rate decreases.

To avoid these and other factors which affect rates of reaction, another approach is often employed. The principle involved is indicated in Figure 4-6. Here, the rate of reaction (activity) of the enzyme is directly related to the amount of active enzyme present when the substrate concentration is in excess. A measure of reaction rate is then a measure of the amount of active enzyme or, better, enzymatic activity. To eliminate the rate change difficulty that can occur during a long incubation, the rate measurements are generally made during the first few minutes of the reaction. This technique is referred to as the initial rate or "kinetic" method.

Perhaps the initial rate technique can be more easily understood if the measurement of lactate dehydrogenase (LD) activity in serum is considered.

113

$$CH_3-\overset{O}{\underset{\|}{C}}-COOH + NAD \cdot H + H^+ \underset{}{\overset{LD}{\rightleftharpoons}} CH_3-\underset{OH}{\underset{|}{CH}}-COOH + NAD^+$$

pyruvic acid   lactic acid

In the reaction, pyruvate is reduced to lactate as coenzyme NAD · H is oxidized. NAD · H absorbs light at a wavelength of 340 nm, whereas NAD$^+$ has a very low absorbance at this wavelength. The amount of light absorbed by NAD · H is directly related to its concentration. Therefore, changes in absorbance provide measurements of concentration changes.

To measure serum LD activity, a mixture of serum, NAD · H, and pyruvate are incubated under conditions of constant temperature and pH. The changes in absorbance at 340 nm are recorded with respect to time by making measurements every 30 seconds or by the use of a strip chart recorder. The results for this example are:

114

Seconds	Δ Absorbance	Δ μmoles / minute
0-30	0.024	0.012
30-60	0.026	0.013
60-90	0.027	0.013
90-120	0.028	0.011
	Average	0.012

Under these conditions, the average initial rate for this enzyme is 0.012 μmoles/minute. The enzyme is converting 12 nmoles of pyruvate to lactate each minute and 12 nmoles of $NAD \cdot H$ to NAD^+ each minute.

Kinetic or initial rate methods are particularly useful for the measurement of enzymes which catalyze reactions that are readily reversible, such as LD.

The two methods, two point and kinetic, provide the same type of data: change in concentration with respect to time, or a rate. Whether the initial rate or steady state technique is used to measure the activity of a given enzyme depends upon several factors. As indicated above, if the reaction rate changes soon after the start of a reaction, the initial rate measurement is preferred. However, in some instances analytic methods are not available to accurately measure the small changes in concentration afforded by short-term reactions so that longer incubations, as for steady state, are required. Instrumentation is also a factor. For the above example, if a photometer for absorbance measurements at 340 nm was not available, then the assay would probably have to be performed by a steady state technique. The solubility of reactants can be a decisive factor. Substrates are sometimes so insoluble that not enough can be dissolved to maintain the concentration in excess for a long incubation. In this instance, an initial rate measurement would be a necessity.

In general, initial rate measurements are preferable. This approach eliminates the several difficulties illuminated above, and with proper instrumentation it allows measurements to be made much faster than most steady state techniques. Initial rate techniques are not limited to absorbance changes at 340 nm as in the above example but can utilize any of the common analytic approaches: titrimetry, colorimetry, fluorometry, and so on.

ACTIVITY UNITS

In order to have meaningful rates it is necessary to consider other factors. One is the enzyme concentration used. If the enzyme is crystalline or fulfills other criteria of purity, the rate can be expressed on a weight basis; micromoles of substrate consumed or micromoles of product formed per minute per milligram, or, as it is usually written, μmoles/minute/mg. However, this is not applicable in clinical chemistry because the weight of the enzyme in the samples is not known. Thus the dimensions used are based on volume: μmoles/ml/minute or μmoles/liter/minute.

A variety of different dimensions have been used in the past for reporting enzymatic activity, e.g., mg/ml/minute, mg/liter/30 minutes, mg/10 ml/30 minutes, and mg/dl/hour. In an effort to eliminate the confusion involved in reporting activity, various professional groups have advocated a system of uniformity. The dimensions proposed are the so-called international units, which are μmoles/ml/minute, μmoles/liter/minute, or μmoles/mg/minute. It is obvious that a consistent unit will be helpful but will not be the complete answer, since many factors affect activity. To make valid activity comparisons, temperature, pH, time, and concentration of all substances in the reaction medium as well as the technique, must be consistent.

NOMENCLATURE

When enzymes such as pepsin, ptyalin, trypsin, and chymotrypsin were first discovered their names did not follow a general pattern of nomenclature. At present, enzymes are named by taking the name of the principle substrate and adding the suffix ase. An example of this would be the term sucrase for an enzyme which catalyzes a reaction with sucrose.

One of the complications which arises from naming enzymes according to their substrate is the fact that many enzymes have several substrates, whereas other enzymes only have one or two substrates. As indicated previously, enzymes have varying degrees of specificity; some can catalyze reactions with only one or possibly two substrates. The problem is further compounded in that enzymes from different sources which catalyze the same reaction may show different specificities. An example of this is the reaction catalyzed by cholinesterases.

115

$$\text{CH}_3-\overset{\overset{\displaystyle O}{\|}}{\text{C}}-\text{O}-\text{CH}_2-\text{CH}_2-\text{N}^+(\text{CH}_3)_3 + \text{H}_2\text{O} \longrightarrow$$
acetylcholine

$$\text{CH}_3-\text{COOH} + \text{HOCH}_2-\text{CH}_2-\text{N}^+(\text{CH}_3)_3$$
acetic acid   choline

Some cholinesterases are almost restricted to actylcholine as a substrate because they show the greatest activity with this substrate. However, there are other cholinesterases, isolated from different tissues in the same organism, that hydrolyze acetylcholine as well as a number of similar substrates (butrylcholine and benzoylcholine) at about the same rate.

Some other esterases with little specificity are able to hydrolyze a number of different esters. They can hydrolyze esters with short-chain alcohol and acid groups or long-chain acid and alcohol groups.

Sometimes the specificity of an enzyme is determined by the configuration of the substrate molecule. Some enzymes will catalyze reactions with L amino acids but not the D isomers. Some enzymes not only show a wide range of specificity but some even show a difference in the type of reaction catalyzed. An example of this fact is that some of the enzymes that are proteolytic enzymes are also able to function to a limited extent as esterases. This discussion of specificity can probably be best summarized by saying that various enzymes show a wide spectrum of specificity. Some are extremely specific and many others seem to be quite nonspecific. This also justifies classifying enzymes by other systems which do not depend solely upon substrates.

CLASSIFICATION

Apparently, as everything else, enzymes can be classified. As indicated for some other things, the number of classification schemes seems to be at least equal to the number of textbook writers or workers in the field or both.

The enzymes have been classified according to the types of protein, thus having some relationship to protein classification schemes. They have been classified according to the components required for activity, according to the type of substrate they attack, and according to the type of chemical reaction carried out. Recently, enzymes have been classified according to a numerical system. Humans are classified by their social security numbers, their telephone numbers, and their zip codes, and now one can also identify enzymes by numbers.

One classification scheme based on the type of reaction is presented in Table 4-1. The system outlined here classifies the enzymes according to the type of reaction they carry out, regardless of the requirement for any cofactors, the type of protein, or the source of the enzyme. No attempt is made to defend this classification scheme as being any way superior or indeed any more or less useful than any other scheme. It is presented merely to indicate to the student the types of chemical reactions that enzymes can carry out and to list a few examples of specific enzymes which will be encountered later.

HYDROLASES

The first classification shown is that of the hydrolases. These enzymes catalyze reactions which result in the decomposition of the substrate by the addition of water.

TABLE 4-1. ENZYME CLASSIFICATION

Enzyme group	Examples
Hydrolases	
Carbohydrase	Sucrase
	Amylase
Esterases	Lipase
	Phosphatase
	Cholinesterase
Proteases	Pepsin
	Trypsin
	Chymotrypsin
	Carboxypeptidase
	Aminopeptidase
Nucleases	Ribonuclease (RNase)
	Deoxyribonuclease (DNase)
Amidases	Glutaminase
	Urease
Transferases	
Transaminase	Glutamate oxalacetate transaminase (GOT)
Transphosphorylase	Phosphoenolpyruvate transphosphorylase
Transglycosidase	Phosphorylase
Transmethylase	Guanidoacetic acid methyltransferase
Oxidoreductases	
Dehydrogenases	Lactate dehydrogenase (LD)
Dehydrases	Malic dehydrase
	Carbonic anhydrase
Peroxidase	
Calatase	

116

(a) $R-\overset{\overset{O}{\|}}{C}-O-R' + H_2O \longrightarrow R-\overset{\overset{O}{\|}}{C}-OH + R'OH$

(b) $NH_2-CHR-\overset{\overset{O}{\|}}{C}-NH-CHR'-COOH + H_2O \longrightarrow$
$NH_2-CHR-COOH + NH_2-CHR'-COOH$

The simple hydrolysis of an ester yielding the corresponding acid and alcohol is shown in **116** (*a*). This type of reaction is carried out by the esterases such as lipase, phosphatase, and cholinesterase. Example (*b*) represents the hydrolysis of a dipeptide to yield two corresponding amino acids. This type of chemical reaction is common for the proteases, or protein-hydrolyzing enzymes.

Carbohydrases are the enzymes which split the glycoside bonds by the addition of water. Examples are:

1. Sucrase: hydrolyzes sucrose to glucose and fructose
2. Maltase: hydrolyzes maltose to 2 moles of glucose
3. Lactase: hydrolyzes lactose to galactose and glucose.

The nucleases are enzymes which hydrolyze nucleic acids and their constituent nucleotides.

Amidases, as would be gathered from their name, hydrolyze amides. Urease is a fine example of an amidase for it hydrolyzes urea or carbamide to yield carbonic acid and ammonia.

TRANSFERASES

The second category is transferases. There are numerous enzymes which fall into this class. Of considerable interest and importance in clinical chemistry are the transaminases which are enzymes that are able to transfer the amino group from an amino acid to a keto acid.

117

$$\begin{array}{c} COOH \\ | \\ H-C-NH_2 \\ | \\ R \end{array} + \begin{array}{c} COOH \\ | \\ C=O \\ | \\ R' \end{array} \longrightarrow \begin{array}{c} COOH \\ | \\ C=O \\ | \\ R \end{array} + \begin{array}{c} COOH \\ | \\ H-C-NH_2 \\ | \\ R' \end{array}$$

This type of reaction will be emphasized later.

Transphosphorylation reactions are those which involve the transfer of a phosphate group from one substance to another and transglycosidation reactions involve the transfer of a carbohydrate moiety from one group to another.

From the foregoing, it is apparent that transmethylation involves transfer of methyl groups from a donor to a recipient. An example of transmethylation will be encountered in the discussion of creatine.

The transferase enzymes perhaps can be summarized as

118 $$D - X + R \rightleftharpoons D + R - X$$

DX represents the donor with X the group to be transferred and R represents the recipient molecule. In the reaction, DX contributes the X group to R to give the products $D + RX$. The group transferred, of course, depends upon the particular enzyme system. For the transaminases, the X group would be an amino group; for the transphosphorylases the group would be a phosphate, and so forth.

OXIDOREDUCTASES

In the oxidoreductases, the first group to be considered is the dehydrogenases. A vast majority of biologic oxidations are carried out by dehydrogenation. An example of this is equation **119**, where lactic acid is oxidized by lactate dehydrogenase (LD) to pyruvic acid by the removal of two hydrogens and two electrons.

119
$$CH_3-\underset{OH}{CH}-COOH \xrightarrow{LD} CH_3-\underset{O}{\overset{\|}{C}}-COOH + 2H\cdot$$

$$\text{lactic acid} \qquad\qquad \text{pyruvic acid}$$

Oxidation can also be accomplished by dehydration.

120
$$\underset{\substack{|\\COOH\\\text{malic acid}}}{\overset{\substack{COOH\\|}}{H-C-OH}}\;\underset{CH_2}{|}\xrightarrow{\text{malic dehydrase}} \underset{CH-COOH}{\overset{HOOC-CH}{\|}} + H_2O$$

$$\text{fumaric acid}$$

Here, malic acid is oxidized to fumaric acid by the removal of water, a reaction catalyzed by malic dehydrase. The remaining two groups of this classification will not be considered.

REVERSIBILITY OF ENZYME REACTIONS

All chemical reactions must be considered reversible even though in some cases we may not be able to demonstrate the reversibility. Generally, this is true when the energy conditions are such that the formation of product is favored so much over the formation of reactants that it cannot be shown with the usual techniques. If an enzyme is to act as a true catalyst and allow the reaction to reach equilibrium sooner than it would without the catalyst, then an enzyme must catalyze the reaction in both directions. This is illustrated very nicely by the action of carbonic anhydrase.

121 $$CO_2 + H_2O \xrightarrow[\text{(at the tissues)}]{\text{carbonic anhydrase}} H_2CO_3$$

$$H_2CO_3 \xrightarrow[\text{(at the lungs)}]{\text{carbonic anhydrase}} H_2O + CO_2$$

Red blood cells have considerable carbonic anhydrase activity. At the tissue level the enzyme catalyzes the hydration of carbon dioxide (CO_2) to form carbonic acid. The same enzyme, in the same cells at the lungs, catalyzes the reaction in the reverse direction; that is, it promotes the decomposition of carbonic acid to water and CO_2. The direction of this particular reaction depends upon the relative concentration of the products and reactants, as for any other reaction. At the tissue level the concentration of CO_2 is high; thus the reaction promoting the formation of carbonic acid is favored. At the lungs, however, the concentration of CO_2 is low and the decomposition of carbonic acid is favored. The reversibility of enzyme-catalyzed reactions is encountered frequently in metabolic reactions.

ENZYME INHIBITION

There are a number of substances which inhibit enzyme activity. The inhibition of enzymes can be divided into two broad categories. One is noncompetitive inhibition, which refers to the reaction of the enzymes with substances (inhibitors) which convert them to inactive forms.

122 $$E + I \longrightarrow EI$$

Here, the enzyme E combines with the inhibitor I to form EI, an inactive enzyme-inhibitor complex. There are many enzymes which require a free thiol group for activity. These enzymes are almost invariably inhibited by heavy metals. For example,

123 $$2\,Pr\!-\!SH + Hg^{++} \longrightarrow (PrS^-)_2Hg + 2\,H^+$$

Two moles of the enzyme react with the mercuric ion to form a complex which is no longer active as an enzyme. Urease is an example of an enzyme which experiences heavy-metal inhibition.

Another type of inhibition is the competitive type, which usually occurs as a result of the enzyme getting confused about what substance is the substrate.

124 (a)
$$\underset{\text{malonic acid}}{\begin{array}{c}\text{COOH}\\|\\\text{CH}_2\\|\\\text{CH}_2\\|\\\text{COOH}\end{array}} \xrightarrow{\text{succinate dehydrogenase}} \underset{\text{fumaric acid}}{\begin{array}{c}\text{HOOC}-\text{CH}\\\|\\\text{CH}-\text{COOH}\end{array}} + 2\text{ H}\cdot$$

$$\begin{array}{c}\text{COOH}\\|\\\text{CH}_2\\|\\\text{COOH}\end{array}$$

(b) $E + C \longrightarrow EC$

$E + S \longrightarrow ES \longrightarrow E + P$

The enzyme succinate dehydrogenase ordinarily catalyzes the oxidation of succinic acid to fumaric acid by the removal of the two hydrogens and two electrons. Malonic acid is a competitive inhibitor of succinic dehydrogenase; malonic acid can combine with the enzyme very much like succinic acid, but it cannot be decomposed to form the fumaric acid product. Thus, the enzyme becomes tied up with malonic acid and is not free to react with the true substrate, succinic acid. As shown in **124** (b), the enzyme E can combine with the substrate to form the reactive complex, which then decomposes to enzyme and product; or the enzyme can combine with C, the competitive inhibitor to form an inactive complex.

Competitive inhibition differs from noncompetitive inhibition in several respects. One is that the substrate and the inhibitor apparently compete for the same catalytic or active site on the enzyme. The competitive inhibitor is bound to the active site and thus prevents the substrate from reaching it. Noncompetitive inhibition does not usually involve the catalytic site. In **123** the thiol group need not be at the catalytic site but may be in some other location on the protein. Competitive inhibition usually can be distinguished from noncompetitive since competitive inhibition can be diminished by increasing the substrate concentration. In a sense, this would amount to diluting out the inhibitor, thus increasing the chances of the enzyme combining with the substrate rather than the competitive inhibitor.

Some enzymes are produced by cells in an inactive form and they must be activated before they will show enzymatic activity. Few of the proteolytic enzymes are active when secreted in the digestive tract until certain chemical changes occur to render them active. These will be discussed in greater detail in Chapter 6.

CHAPTER 5

NUCLEIC ACIDS

In recent years, the function of nucleic acids in the transmission of genetic characteristics has been elucidated. The nucleic acids serve as patterns or templates for the synthesis of proteins and consequently are of fundamental importance in genetics. Nucleic acids are polymers that occur in cells usually associated with proteins.

PYRIMIDINE AND PURINE BASES

The primary variable elements of the nucleic acids are the purine and pyrimidine bases. Pyrimidine bases consist mainly of uracil, cytosine, and thymine.

125

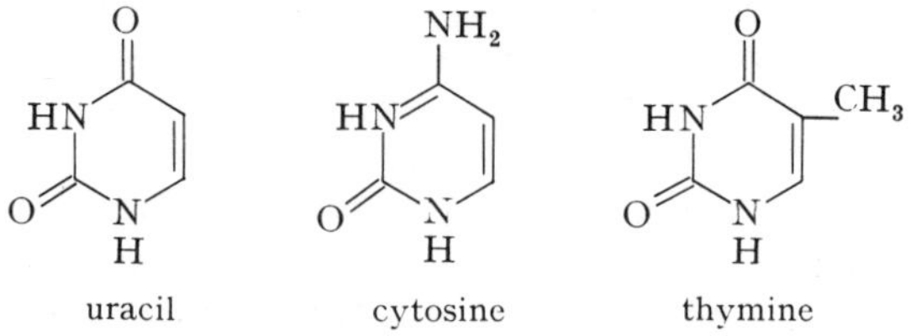

uracil   cytosine   thymine

The numbering system for the pyrimidine ring is seen in **126**.

126

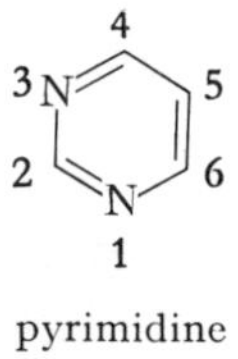

pyrimidine

Using this numbering system, uracil would be 2,4-dioxopyrimidine; cytosine would be 2-oxo-4-aminopyrimidine, and thymine would be 2,4-dioxo-5-methylpyrimidine. There are other pyrimidine bases which occur naturally in nucleic acids, but their concentrations are minor.

The purine bases which occur predominantly in nucleic acids are adenine and guanine.

127

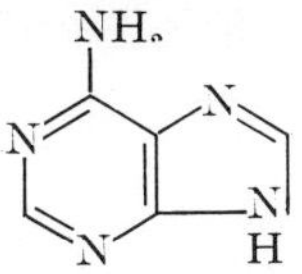

adenine 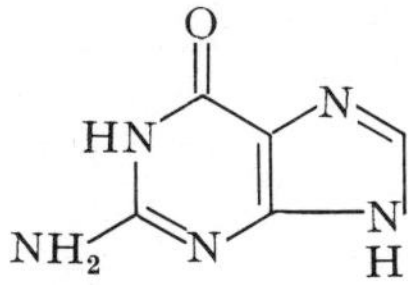   guanine

The purine numbering system is as follows:

128

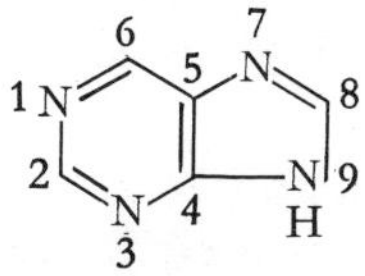

purine

From this it can be realized that adenine could be called 6-amino purine; and guanine, 2-amino-6-oxopurine.

NUCLEOSIDES

The bases combine with ribose to yield nucleosides. A pyrimidine nucleoside is uridine, which has β-D-ribose connected to the number 1 nitrogen of uracil.

129

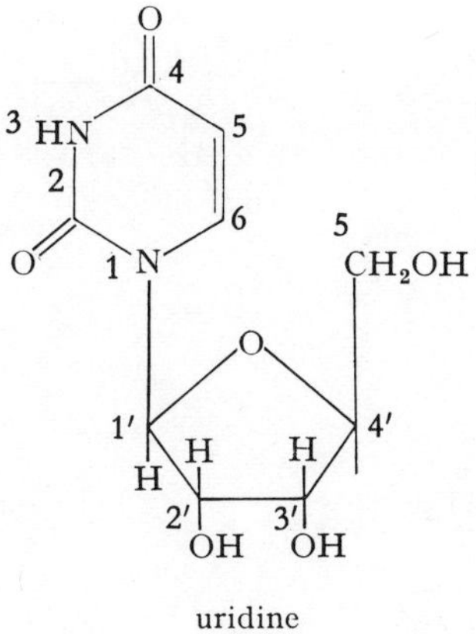

uridine
(1-β-D-ribofuranosyluracil)

Another nucleoside, deoxyadenosine, has 2-deoxyribose connected to the number 9 position of adenine.

130

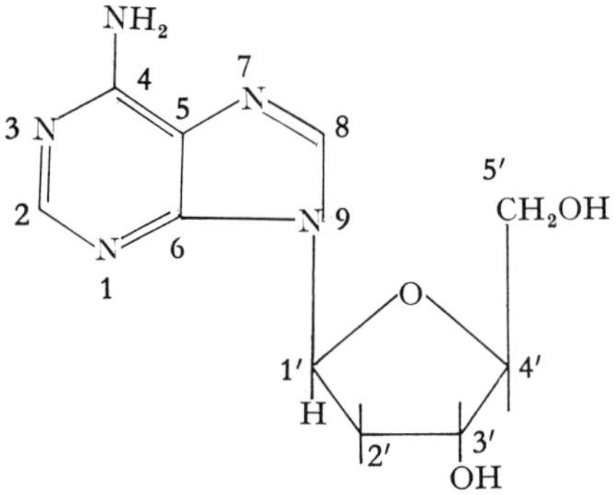

deoxyadenosine
(9-β-D-2'-deoxyribofuranosyladenine)

All three of the pyrimidines and both purine bases can combine with ribose or deoxyribose as indicated in **129** and **130**.

NUCLEOTIDES

The nucleosides can form monophosphate esters to become nucleotides.

131

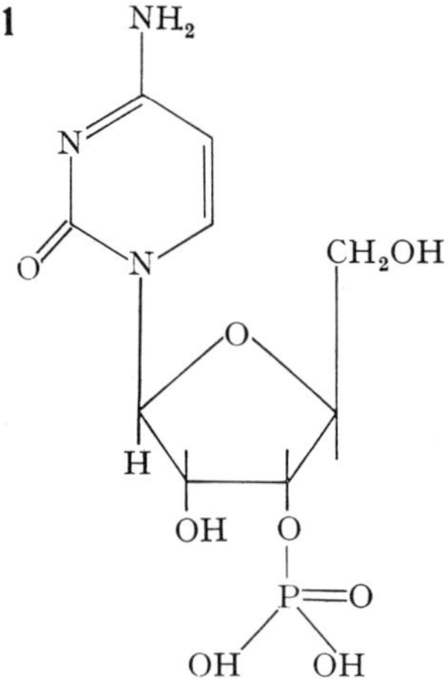

3'-cytidylic acid
(cytidine-3'-phosphate)

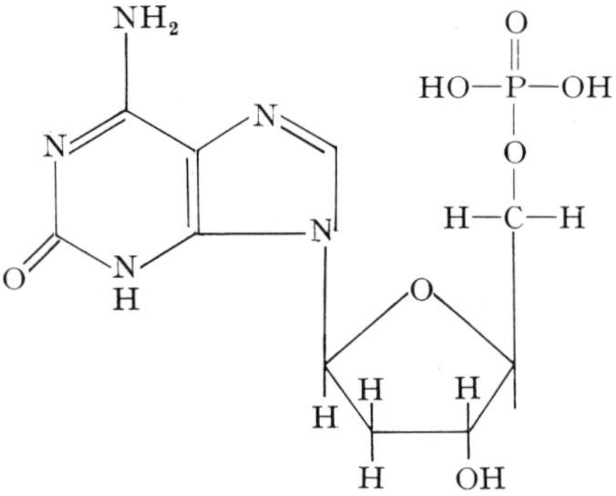

5'-deoxyguanylic acid
(deoxyguanosine-5'-phosphate)

TABLE 5-1. DERIVATIVES OF THE PYRIMIDINE AND PURINE BASES

Base	Nucleoside	Deoxynucleoside	Nucleotide	Deoxynucleotide
Uracil	Uridine	Deoxyuridine*	Uridylic acid (uridine-3' or 5'-phosphate)	Deoxyuridylic acid* (dexoyuridine-3' or 5'-phosphate)
Cytosine	Cytidine	Deoxycytidine	Cytidylic acid (cytidine-3' or 5'-phosphate)	Deoxycytidylic acid (deoxycytidine-3' or 5'-phosphate)
Thymine	Thymidine†	Deoxythymidine†	Thymidylic acid† (thymidine-3' or 5'-phosphate)	Deoxythymidylic acid (deoxythymidine-3' or 5'-phosphate)
Adenine	Adenosine	Deoxyadenosine	Adenylic acid (adenosine-3' or 5'-phosphate)	Deoxyadenylic acid (deoxyadenosine-3' or 5'-phosphate)
Guanine	Guanosine	Deoxyguanosine	Guanylic acid (guanosine-3' or 5'-phosphate)	Deoxyguanylic acid (deoxyguanosine-3' or 5'-phosphate)

* Does not normally occur in nucleic acids or occurs only as a minor component.
† When thymine nucleoside and nucleotide were first named, they were thought to occur only as the deoxyriboside and deoxyribotide; the name thymidine was assigned to the former and thymidylic acid to the latter. To surmount this difficulty, the names ribothymidine and ribothymidylic acid are often used to describe the nucleoside and nucleotide, respectively, of thymine and ribose.

On the left is the 3'-phosphate ester of cytidine called 3'-cytidylic acid or cytidine-3'-phosphate. On the right is the 5'-phosphate ester of deoxyguanosine, sometimes called 5'-deoxyguanylic acid. Refer to Table 5-1, for the name of various nucleosides and nucleotides.

Purines and pyrimidines are spoken of as bases, but in reality they are weak acids. The pyrimidine base uracil **132** can undergo ketoenol tautomerism and can ionize to release a proton from either the keto or enol form. The term oxo is frequently used to describe the keto or enol functional group on purines and pyrimidines.

132

POLYNUCLEOTIDE STRUCTURES

The following trinucleotide structure represents the type of diester linkage found in the nucleic acids.

133

This particular structure contains ribonucleotides, but the structure of the deoxyribonucleotide polymers is very similar in that the diester linkage connecting the nucleotides occurs between the 3' and 5' positions. Using these abbreviations: bases, B; ribose, R;

deoxyribose, dR; phosphate, (P), a polynucleotide can be represented as

134

$$\underset{(P)}{}-\underset{\overset{|}{R}}{\overset{B}{}}-\underset{(P)}{}-\underset{\overset{|}{R}}{\overset{B}{}}-\underset{(P)}{}-\underset{\overset{|}{R}}{\overset{B}{}}-\underset{(P)}{}-\underset{\overset{|}{R}}{\overset{B}{}}\ldots$$

or

$$(P)-\overset{\overset{B}{|}}{R}-(P)-\left(\overset{\overset{B}{|}}{R}-(P)\right)_{x}-\overset{\overset{B}{|}}{R}$$

If similar brevity is used, a polydeoxynucleotide can be represented as

135

$$(P)-\overset{\overset{B}{|}}{dR}-(P)-\left(\overset{\overset{B}{|}}{dR}-(P)\right)_{x}-\overset{\overset{B}{|}}{dR}$$

The occurrence of the bases in various nucleic acids is not fixed but depends upon the source of the nucleic acids. If we represent the various bases by the first letter of their names—uracil, U; thymine, T; and so on—a tetranucleotide composed of adenylic, uridylic, guanylic, and cytidylic acids can be represented in this manner.

136

$$(P)-\overset{\overset{A}{|}}{R}-(P)-\overset{\overset{U}{|}}{R}-(P)-\overset{\overset{G}{|}}{R}-(P)-\overset{\overset{C}{|}}{R}$$

The nucleic acids can be separated into two types. One type, known as ribonucleic acid (RNA), is characterized by nucleotides with ribose. A second type of nucleic acid is called deoxyribonucleic acid (DNA) and is characterized by having deoxyribose in the molecule. These structures, RNA and DNA, are represented in **134** and **135**, respectively. As a generalization, RNA contains the bases U, C, A, and G; and DNA has T, C, A, and G.

NUCLEOTIDE DERIVATIVES

Adenosine triphosphate (ATP), a nucleotide derivative, is important in metabolism.

137

$$\text{adenine-ribose-}CH_2-O-\overset{\overset{O}{\|}}{\underset{OH}{P}}-O-\overset{\overset{O}{\|}}{\underset{OH}{P}}-O-\overset{\overset{O}{\|}}{\underset{OH}{P}}-OH$$

ATP
(adenosine triphosphate)

This metabolite contains bonds designated as "high-energy" bonds. These bonds are frequently represented as lightning bolts.

138

$$\text{adenosine}-O-\overset{\overset{O}{\|}}{\underset{OH}{P}}-O\sim\overset{\overset{O}{\|}}{\underset{OH}{P}}-O\sim\overset{\overset{O}{\|}}{\underset{OH}{P}}-OH$$

ATP

In biochemical parlance, the term high energy pertains to compounds capable of releasing large amounts of energy when the bond is broken.

Adenosine triphosphate in an aqueous biologic system near neutrality exists as a salt and actually should be represented as such.

139

$$\text{adenosine}-O-\overset{\overset{O}{\|}}{\underset{O^-Na^+}{P}}-O\sim\overset{\overset{O}{\|}}{\underset{O^-Na^+}{P}}-O\sim\overset{\overset{O}{\|}}{\underset{O^-Na^+}{P}}-O^-Na^+$$

ATP

The cations, sodium in this example, are free to migrate, but the charges on the phosphate groups are fixed. As indicated in the discussion of hydrogen bonds, oxygen has a great attraction for electrons. The electrons in the bond, P=O, shift toward the oxygen leading to polarization. The structure of ATP using $\delta(-)$ for polarization, would be

140

$$\text{adenosine}-\overset{\overset{\delta(-)O}{\|}}{\underset{O^-}{P}}-O\sim\overset{\overset{\delta(-)O}{\|}}{\underset{O^-}{P}}-O\sim\overset{\overset{\delta(-)O}{\|}}{\underset{O^-}{P}}-O^-$$

There is an accumulation of electrostatic negative charges, $-$, and negative polarization, $\delta(-)$. Since the centers of negativity repel one another these bonds are extremely reactive and unstable.

Phosphotransferase enzymes can catalyze the transfer of the phosphoryl group along with much of the energy associated with the high-energy bond.

141

$$\text{adenosine} -\text{O}-\underset{\underset{\text{OH}}{|}}{\overset{\overset{\text{O}}{\|}}{\text{P}}}-\text{O} \sim \underset{\underset{\text{OH}}{|}}{\overset{\overset{\text{O}}{\|}}{\text{P}}}-\text{O} \sim \underset{\underset{\text{OH}}{|}}{\overset{\overset{\text{O}}{\|}}{\text{P}}}-\text{OH} + \text{R}-\text{OH} \longrightarrow$$

ATP

$$\text{adenosine} -\text{O}-\underset{\underset{\text{OH}}{|}}{\overset{\overset{\text{O}}{\|}}{\text{P}}}-\text{O} \sim \underset{\underset{\text{OH}}{|}}{\overset{\overset{\text{O}}{\|}}{\text{P}}}-\text{OH} + \text{R}-\text{O}-\underset{\underset{\text{OH}}{|}}{\overset{\overset{\text{O}}{\|}}{\text{P}}}-\text{OH}$$

ADP (adenosine diphosphate)

With the transfer of energy, the recipient molecules become reactive if the energy levels are sufficiently increased. Energy transfer is important in metabolism, because the recipient molecules become reactive under the mild conditions compatible with life processes. Examples will be presented in succeeding chapters.

Polyphosphate derivatives of the other nucleotides can also be formed. Refer to Table 5-2 for their names and usual abbreviations.

TABLE 5-2. POLYPHOSPHATE DERIVATIVES: NAMES AND ABBREVIATIONS

Monophosphate	Diphosphate	Triphosphate
5'-Uridylic acid 5'-Uridine monophosphate (UMP)	Uridine diphosphate (UDP)	Uridine triphosphate (UTP)
5'-Cytidylic acid 5'-Cytidine monophosphate (CMP)	Cytidine diphosphate (CDP)	Cytidine triphosphate (CTP)
5'-Adenylic acid 5'-Adenosine monophosphate (AMP)	Adenosine diphosphate (ADP)	Adenosine triphosphate (ATP)
5'-Guanylic acid 5'-Guanosine monophosphate (GMP)	Guanosine diphosphate (GDP)	Guanosine triphosphate (GTP)

Deoxyribotides form similar derivatives. The names are the same except for the prefix deoxy (e.g., 5'-deoxyadenosine monophosphate or 5'-deoxyadenylic acid). Abbreviations include 5' dGMP = deoxyguanylic acid = 5'-deoxyguanosine monophosphate.

Nucleic acids are formed by the condensation of triphosphates and the elimination of pyrophosphate. Using the abbreviations presented earlier, polymerization can be represented as

142 $\quad n\ B-R-\!\!\text{\large (}\!P\!\text{\large)}\!-\!\text{\large (}\!P\!\text{\large)}\!-\!\text{\large (}\!P\!\text{\large)} \longrightarrow$

$$B-R-\text{(P)}-\left(\begin{array}{c}B\\|\\R-\text{(P)}\end{array}\right)_{n-2}\!\!\!-\!\!\begin{array}{c}B\\|\\R\end{array}\!-\!\text{(P)} \ +\ n\ HO-\overset{\overset{O}{\|}}{\underset{\underset{OH}{|}}{P}}-O-\overset{\overset{O}{\|}}{\underset{\underset{OH}{|}}{P}}-OH$$

pyrophosphate

Many interesting metabolic intermediates are formed from the nucleotide triphosphates. For example, uridine triphosphate reacts with glucose-1-phosphate to form uridine diphosphoglucose.

143 $\qquad\text{UTP} + \text{glucose}-1-\text{(P)} \longrightarrow$

uridine—(P)—(P)—O—[glucose ring with CH₂OH, OH, OH, OH]

uridine diphosphoglucose
(UDP-glucose)

This nucleotide derivative functions in the conversion of galactose to glucose.

Adenosine triphosphate forms an acid anhydride with amino acids.

144

$$\text{ATP} + \underset{\underset{NH_2}{|}}{R-CH}-COOH \longrightarrow P-P_i + \text{adenosine}-O-\overset{\overset{O}{\|}}{\underset{\underset{OH}{|}}{P}}-O\sim\overset{\overset{O}{\|}}{\underset{\underset{NH_2}{|}}{C}}-CHR-R$$

pyrophosphate   aminoacyl adenylate

The formation of these anhydrides provides the amino acid residues with sufficient energy so that they can participate in protein synthesis. This reaction is frequently referred to as amino acid activation.

Several coenzymes are nucleotide derivatives. The NAD^+ molecule is a dinucleotide.

145

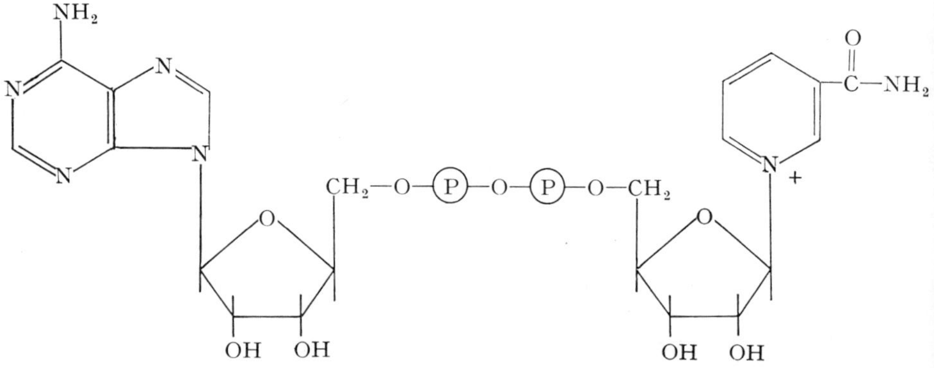

nicotinamide adenine dinucleotide NAD^+

A reaction discussed in the chapter on enzymes shows this coenzyme is reduced as the substrate is oxidized. A portion of this reaction is

146 $\qquad NAD^+ + 2H\cdot \rightleftarrows NAD\cdot H + H^+$

In the NAD^+ molecule the portion that undergoes change in oxidation state is nicotinamide, a water soluble vitamin.

147

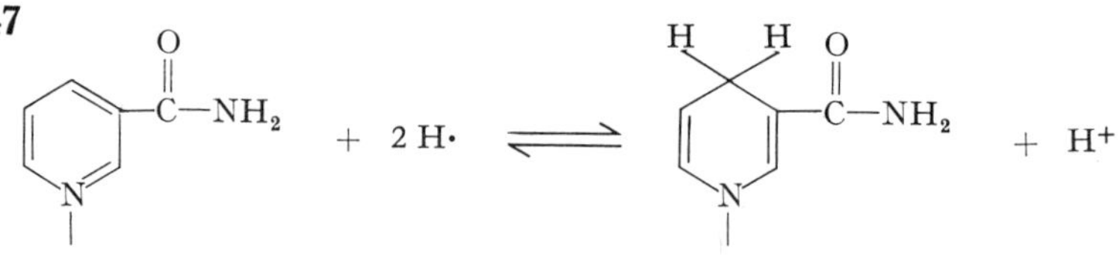

$NAD^+ \qquad\qquad NAD\cdot H$

The coenzyme $NADP^+$ is similar in structure and function to NAD^+. However, in $NADP^+$ the ribose attached to adenine has an additional phosphate ester in the 2' position. The abbreviation $NADP^+$ originates from nicotinamide adenine dinucleotide phosphate.

CHAPTER 6

DIGESTION

Very few foods ordinarily eaten are of any value to the organism in the state in which they are consumed. There are exceptions, however, in that water, some small saccharides, vitamins, and inorganic salts can be used without further processing. Also, many drugs can be absorbed and utilized without digestive processes. Most foodstuffs must undergo a series of changes before they can be absorbed and utilized. The subject of digestion covers all the processes which take place that make foodstuffs absorbable and utilizable by the organism.

The cooking of food is not absolutely necessary prior to digestion, but it does help since a great number of substances are partially hydrolyzed and protein is denatured. A side advantage of heating food prior to ingestion is that the process destroys parasites and bacteria.

MOUTH

Digestion begins in the mouth. Food is broken down into small particles by the process of mastication and at the same time is mixed with saliva. The function of saliva is twofold. First, it moistens and lubricates food so that it can be more easily swallowed. In addition, it contains a carbohydrate-digesting enzyme called amylase. Saliva is produced by the parotid, submaxillary, sublingual and the buccal glands. A saliva output of some 1,500 ml per day is considered average for an adult. The secretion of saliva is largely under the control of the nervous system. A variety of stimuli will evoke the elaboration of saliva. The sight, smell, or just the thought of food, as well as mechanical stimuli, are sufficient to start the flow. The composition of saliva is variable depending on a number of different factors. The pH is about neutral, and the protein content is very low.

The enzyme of saliva, amylase, hydrolyzes amylose or the amylose-like portions of starch molecules to yield maltose and smaller starch fragments. The optimum pH for amylase is about 7. The length of time for amylase activity is limited, because when the food passes from the mouth to the stomach it encounters a very acidic environment, which inhibits further amylase activity.

STOMACH

Gastric digestion consists primarily of proteolytic reactions. The gastric wall secretes gastric juice in response to a number of different stimuli. The thought, smell, and taste of food, as well as the mechanical stimulation brought about by the presence of food in the stomach, are effective stimuli. Gastric juice is characterized by the presence of HCl in a concentration of approximately $0.17\ N$ with a pH of about 0.87. The mechanism whereby the gastric mucosa is able to produce and secrete this very concentrated HCl solution is unknown, although we do know that the proton arises from water and the chloride from the blood chloride. An overall series of reactions leading to the formation of HCl is seen in Figure 6-1.

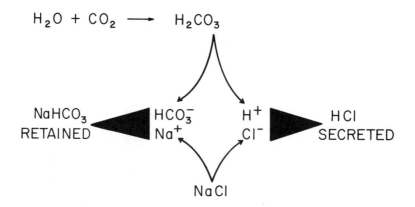

Fig. 6-1. Reaction in the formations of gastric hydrochloric acid.

Obviously in order to secrete a concentrated acid such as this, energy is required. Note that sodium bicarbonate is retained within the vascular system. Since the blood is slightly alkaline in nature, one would expect that it would become more alkaline as sodium bicarbonate was being produced. Indeed this is the case and gives rise to a well-recognized phenomenon often called alkaline tide.

The parietal cells in the gastric wall secrete HCl, and the chief cells secrete a proenzyme called pepsinogen. Pepsinogen is an inactive form of the enzyme pepsin. In the presence of HCl, pepsinogen undergoes a hydrolytic change with the release of several peptides to form the active proteolytic enzyme pepsin.

148 $\quad$ pepsinogen $\xrightarrow{\text{HCl}}$ pepsin + peptides

This reaction is also autocatalytic which means that the product of the reaction, pepsin, is able to convert pepsinogen to pepsin. Pepsin is an endopeptidase, meaning it attacks the interior of molecules and not the terminal amino acids. It hydrolyzes peptide bonds involving the amino groups of tryptophan, phenylalanine, and tyrosine. Pepsin attacks a protein molecule only at the sites indicated and does not hydrolyze a protein completely; it merely breaks it up into smaller peptide units.

ANALYSIS OF GASTRIC CONTENTS

The analysis of gastric contents for HCl and pepsin activity is often used for diagnostic purposes. Usually gastric contents are collected by placing a tube in the stomach and withdrawing specimens at definite intervals. The samples can be basal (that is, those collected without any stimulation) or can be collected following various stimuli. The stimulants commonly used are ethyl alcohol, caffeine, and histamine.

The HCl and other acids in gastric contents can most conveniently be quantitated by titration procedures. Figure 6-2 shows the titration curves for HCl and acetic acids. As can be seen, where pH is plotted against the milliliters of NaOH added, the pH of a HCl solution changes very little until just before the equivalence point, when there is a very fast increase in the pH until the pH becomes alkaline. For organic acids (such as acetic acid) the initial pH is higher and the equivalence point is reached with a gradual change rather then a sharp change. Because of the difference in the behavior of the two types of acids, it is possible to quantitate HCl and strong organic acids when they occur as a mix-

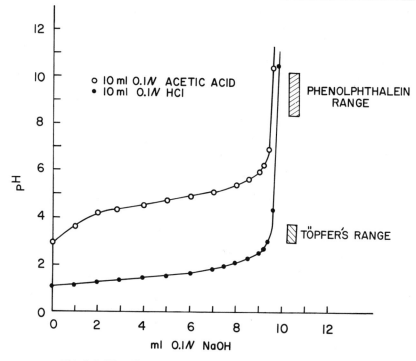

Fig. 6-2. Titration curves for hydrochloric and acetic acids.

ture. Hydrochloric acid can be measured by titrating to pH 3, which is essentially the end point for the HCl titration. If the titration is continued to pH 8 or 9 the total amount of acid present can be quantitated. For the titrimetric analysis of gastric contents this difference in pH at the equivalence points is utilized to separate the titration values for HCl and other organic acids present. To do this Töpfer's reagent (methyl yellow) a pH indicator, is added to the solution and is titrated until the methyl yellow changes color. Methyl yellow is red at pH of about 3 and changes to yellow at pH 4. When the methyl yellow end point is reached, phenolphthalein is added to the sample and the titration continued until the phenolphthalein end point is reached. The amount of NaOH needed to titrate to the methyl yellow end point is then a measure of the HCl present, and the amount of NaOH required to titrate from the beginning to the phenolphthalein end point then is a measure of the total acid present in the sample. Therefore, the number of milliliters required to go from the methyl yellow end point to the phenolphthalein end point is a measure of the organic acids present. A pH meter in place of the indicator can also be used for this analysis.

The normal values for gastric acidity depend on whether the sample is a basal sample or one collected following stimulation. In addition, the normal values vary depending on the length of time the sample is collected. A sample of gastric juice collected under basal conditions from 6:00 P M to 6:00 A M varies from 1 to 90 meq of HCl per liter with an average of about 30. The volume for such samples can range between 140 and 1,200 ml.

Normal individuals usually demonstrate a maximum of HCl secretion approximately 40 minutes after histamine stimulation, and the concentration of HCl averages about 90 meq/liter in a 40-minute sample. The production of gastric juice following histamine stimulation averages about 30 ml per minute.

Achlorhydria is a term used to describe gastric juice which contains no HCl, a condition often associated with pernicious anemia. Hypoacidity or hyposecretion of HCl by the gastric wall is commonly observed in cases of carcinoma of the stomach and less commonly with certain other diseases. Hyperacidity or hypersecretion of gastric juice has long been associated with duodenal ulcers.

If the gastric contents contain HCl, pepsin activity will be present; if it does not, no pepsin activity will be demonstrable. Even though pepsin activity is less frequently analyzed, the absence of pepsin activity confirms achlorhydria.

SMALL INTESTINE

When food leaves the stomach and enters the small intestine, digestion becomes quite complex. The pH of the intestine is near neutrality. In addition to digestion, absorption occurs and the influence of the pancreatic, intestinal wall, and hepatic secretions all come into play. The secretions from the three sources are very effective in neutralizing the high acidity carried from the stomach. The secretions have a pH of about 8 or above and all contain large amounts of sodium bicarbonate, which reacts with the HCl of the stomach.

PANCREATIC SECRETION

Pancreatic juice contains several enzymes which attack macromolecules. Included among these are three proteolytic enzymes. Trypsin and chymotrypsin are endopeptidases elaborated by the pancreas in inactive forms known as trypsinogen and chymotrypsinogen, respectively. An enzyme secreted by the intestinal mucosa attacks trypsinogen, splits off a peptide, and converts it to the proteolytically active trypsin. Chymotrypsinogen is converted to active chymotrypsin in a somewhat complicated process by the action of trypsin and chymotrypsin. Trypsin attacks peptide linkages which are formed by the carboxyl group of either lysine or arginine, and chymotrypsin attacks peptide bonds formed by the carboxyl group of tyrosine, tryptophan, and phenylalanine. The third major proteolytic enzyme from the pancreas is carboxypeptidase. This is a nonspecific enzyme which hydrolyzes peptides, one amino acid at a time, starting from the carboxyl end of the chain.

Pancreatic juice also contains amylase activity. The amylase of pancreatic origin appears to be identical with that of the salivary glands. Amylase attacks starches and hydrolyzes 1,4 links to produce maltose from amylose or straight-chain portions of starch molecules.

Lipase is also found in pancreatic juice. Lipase is an esterase which hydrolyzes triglycerides with long-chain fatty acids first to diglycerides, then monoglycerides; and if the reaction time is long enough, free glycerin is liberated.

Ribonuclease (RNase) in pancreatic juice hydrolyzes RNA to the corresponding nucleotides. DNase (deoxynucleotidase) hydrolyzes deoxynucleic acids to the deoxynucleotide level.

The analysis of pancreatic juice obtained by intubation and collected following stimuli is often used for the detection of pancreatic carcinoma. The analysis of amylase and lipase activities, and bicarbonate content, are common in this procedure.

INTESTINAL WALL SECRETION (SUCCUS ENTERICUS)

The intestinal wall secretes enzymes which attack smaller substrate molecules. Included in this secretion is an enzyme called oligo-1-6-glucosidase. This enzyme is able to split the 1-6 links in starch molecules (branch points) and continue starch hydrolysis. In addition, the intestinal wall secretes the enzymes maltase, which hydrolyzes maltose to glucose; sucrase, which hydrolyzes sucrose to glucose and fructose; and lactase, which hydrolyzes lactose to its constituent monosaccharides.

Enzymes which hydrolyze proteins are also secreted by the wall of the intestines. Aminopeptidase hydrolyzes polypeptides, one amino acid at a time starting from the amino end. In addition, there are other enzymes which complete the hydrolysis of proteins to the constituent amino acids. Also, enzymes which complete the digestion of nucleotides are elaborated by the intestinal wall.

HEPATIC SECRETION (BILE)

Bile probably is produced continuously by the liver and is stored between digestive periods in the gallbladder. Liver bile is green-yellow, viscous and slimy in character, and is alkaline, (pH about 8). Between 500 and 1,000 ml of liver bile is produced each day. While the bile is stored in the bladder it becomes concentrated, since the bladder wall is able to remove water and some of the salts.

Indeed, bile may be considered a unique product. It represents a pathway for the disposal of certain waste products from the organism. It contains no digestive enzymes, and its contribution to digestion consists of certain emulsifying and stimulative agents. The primary constituents of bile and those that will be considered here include bile pigments, bile salts, and cholesterol. Although these three items occur in the greatest concentration, they are not the only constituents. However, discussion of the other substances is beyond the scope of this chapter.

Bile Pigments

Bile pigments consist of derivatives of hemoglobin. When erythrocytes are destroyed in the reticuloendothelial system hemoglobin is decomposed (see **149**).

The protein portion, globin, is removed, and the amino acids are returned to the amino acid pool. Iron is removed and the cyclic tetrapyrrole structure labeled heme is released. The removal of one of the methylene bridges opens the cyclic structure and yields first biliverdin and finally the linear tetrapyrrole compound bilirubin. Bilirubin is then carried to the liver where it is normally conjugated primarily with glucosiduronic acid and then excreted with the bile. Some of the bilirubin is oxidized back to biliverdin, which is secreted with the bile. These two compounds contribute to the color of bile; biliverdin is green and bilirubin is red-orange. When bilirubin enters the intestinal tract it undergoes a number of metabolic changes and is excreted with the feces in the form of urobilinogen. The brown color of fecal material is due primarily to the presence of urobilinogen. Bilirubin apparently does not contribute to the digestive or absorptive processes. The presence of bilirubin and biliverdin in bile simply represents their pathway of elimination.

149

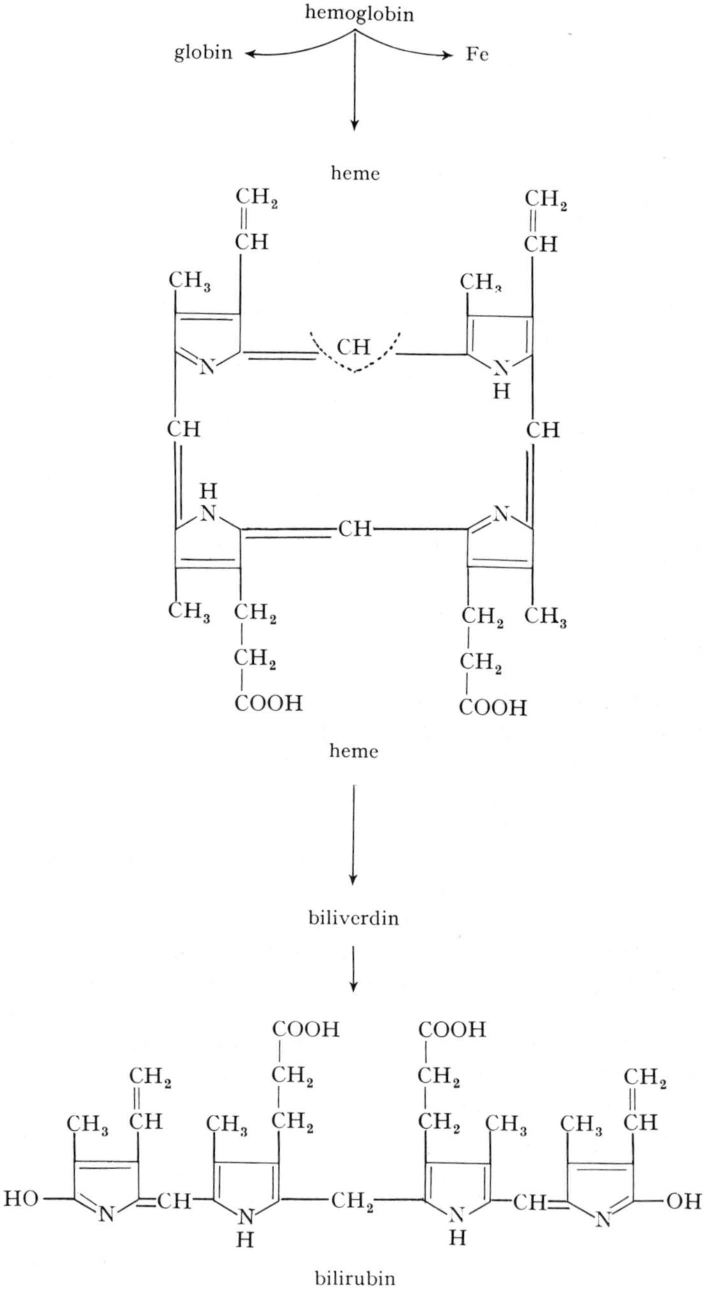

hemoglobin → globin, Fe, heme

heme → biliverdin → bilirubin

Bile salts

The bile salts are cholesterol metabolites. Cholesterol can be degraded to form several acids, one of which is cholic acid (see **150**).

150

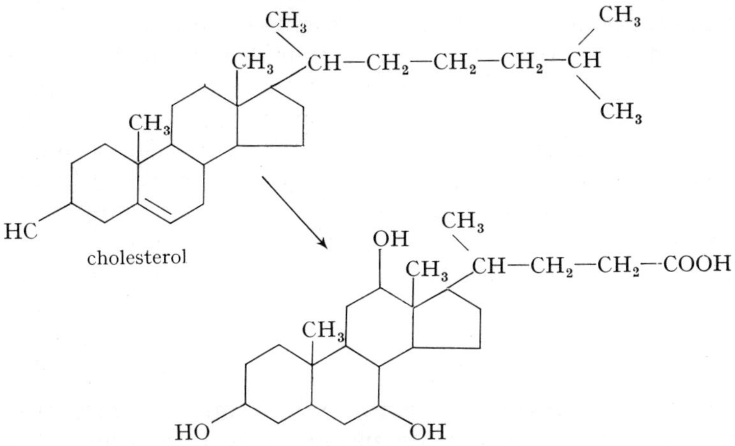

cholesterol

cholic acid

The various bile acids prior to secretion are conjugated with amino acids such as glycine to produce glycocholic acid or with aminoethyl sulfonic acid (taurine) to form taurocholic acid. This is shown in **151** where cholic acid is represented as R—CH$_2$—CH$_2$—COOH

151   R—CH$_2$—CH$_2$—COOH + NH$_2$—CH$_2$—COOH $\xrightarrow{-H_2O}$
           cholic acid   glycine

$$R-CH_2-CH_2-\overset{O}{\underset{\|}{C}}-NH-CH_2-COOH$$
glycocholic acid

R—CH$_2$—CH$_2$—COOH + NH$_2$—CH$_2$—CH$_2$—SO$_3$H $\xrightarrow{-H_2O}$
   cholic acid   taurine

$$R-CH_2-CH_2-\overset{O}{\underset{\|}{C}}-NH-CH_2-CH_2-SO_3H$$
taurocholic acid

These compounds are, of course, secreted as a salt, and it might be of interest to note that the salt of taurocholic acid is indeed a detergent.

The bile acids do not represent the end product of cholesterol metabolism since much cholesterol is eliminated unchanged in the bile.

One of the great difficulties with fat digestion is that fat-splitting enzymes such as lipase are water soluble, whereas fats are water insoluble. The water-soluble enzymes and water-immiscible lipids must come together before enzymolysis can take place. The soap or detergent action of bile salts emulsifies intestinal contents and allows hydrolysis to occur. Through emulsification they also facilitate absorption by the intestinal wall. Included among these compounds are the fat-soluble vitamins D and K. The bile salts are also absorbed and carried to the liver where they stimulate the production of more bile; they themselves are then reexcreted.

Bile contains about 5% cholesterol. The cholesterol, so far as is known, serves no purpose in digestion. The occurrence of cholesterol in the bile probably represents the main excretory route.

Gallstones

In certain individuals, women more commoly than men, the constituents of bile precipitate in the bladder to form gallstones. The stones may occur as a single stone or there may be a multitude of them. If gallstones are examined on cross section, they are found to contain a central nucleus with concentric layers of deposits. Gallstones normally contain cholesterol, the bile pigments, calcium salts (particularly $CaCO_3$), and phosphates. Just why gallstones form in some individuals and not others is not clear, but it apparently is not related to the degree of concentration of the bile while in the bladder.

LARGE INTESTINE

Most of the digestible material is absorbed by the small intestine. Undigested material, unabsorbed substances, cellular debris, and certain vegetable fibers pass from the small intestine into the large intestine. Secretions from the large intestinal wall make the medium slightly alkaline and as a result make conditions ideal for bacterial growth: the temperature is 37º C.; there is no light, little oxygen, and there is ample material for bacterial growth. About one-fourth to one-half of fecal material is made up of bacteria. Although water and some salts are absorbed from the large intestine, the main purpose of this area seems to be as a storage place until the feces can be conveniently evacuated.

ABSORPTION

The preceding material has dealt with the degradation processes whereby complex food substances are degraded to their monomeric or at least smaller and simpler subunits. These digestive processes are necessary before the materials can enter the body. The process whereby these materials are transported from the lumen of the small intestine into the body is called absorption. Not a great deal is known about the absorption of many substances. It is apparently quite complex and rather selective. It is known, however, that energy is required to absorb glucose. The absorption of glucose from the intestine can occur at very low concentration even though the concentration in the body is very high. Thus, the absorption mechanism is not simple diffusion but is often termed an active process, meaning one where work must be done to transport the glucose. The processes involved in the absorption of amino acids and lipids has been and still is the source for considerable controversy, and the mechanisms are not at all clear.

Most of the absorption from the digestive tract takes place from the small intestine. However, apparently ethyl alcohol and certain other substances (some drugs) can be absorbed from the mouth and stomach.

CHAPTER 7

CARBOHYDRATE METABOLISM

The term metabolism can be defined as all the physical and chemical changes that occur within living systems. The concept is easy to grasp if we consider that substances like glucose are absorbed into the organism and oxidized to CO_2 and water. In this case metabolism refers to the degradation of glucose to its end products with concomitant production of energy. The degradative processes are called catabolism. There are also accumulative aspects for which the term anabolism is used. Changes that take place during the growth of plants and animals are anabolic, whereas the wasting away processes are catabolic. In the normal adult animal both processes are balanced to a large degree.

From the foregoing, one could easily conclude that metabolism is the chemical changes which occur to various compounds during their exposure to living systems. Unfortunately, it is also used to describe the transport and compartmentalization of substances which do not undergo chemical change. For instance, many investigators speak of the metabolism of sodium. Sodium enters the organism as a sodium ion, spends a variable period of time in the system, and is released in the same form in which it entered. Obviously, no chemical change has occurred. The term is more often used, however, to describe physical and chemical changes of substances that take place within living systems.

GLYCOLYSIS

Glucose, fructose, and galactose following digestive processes are absorbed and carried by the hepatic portal vein to the liver. Glucose can pass through the liver to other cells of the body or be retained. If it is retained it can be: (1) metabolized to CO_2 and water with the production of energy, (2) stored in the form of glycogen, or (3) metabolized to other products.

When glucose is taken up by the liver, it reacts with adenosine triphosphate to form glucose-6-phosphate and adenosine diphosphate. The terminal phosphate of adenosine triphosphate is transferred to the number 6-hydroxy group of glucose. This reaction is, for practical purposes, irreversible.

152 glucose + ATP $\xrightarrow{\text{glucokinase}}$ glucose-6-(P) + ADP

If glucose is to be converted to glycogen, glucose-6-phosphate must be converted to glucose-1-phosphate.

153 glucose-6-(P) $\xrightleftharpoons{\text{phosphoglucomutase}}$ glucose-1-(P)

Glucose-1-phosphate then reacts with uridine triphosphate to form uridine diphosphate glucose. This reaction was presented in the chapter on nucleic acids (**143**).

154 glucose-6-(P) + UTP $\rightleftharpoons$ UDP-glucose + P-P$_i$

Glycogen is produced by adding one glycosyl unit at a time to the nonreducing terminals of an existing molecule. This is accomplished by reaction of UDP-glucose with glycogen.

155

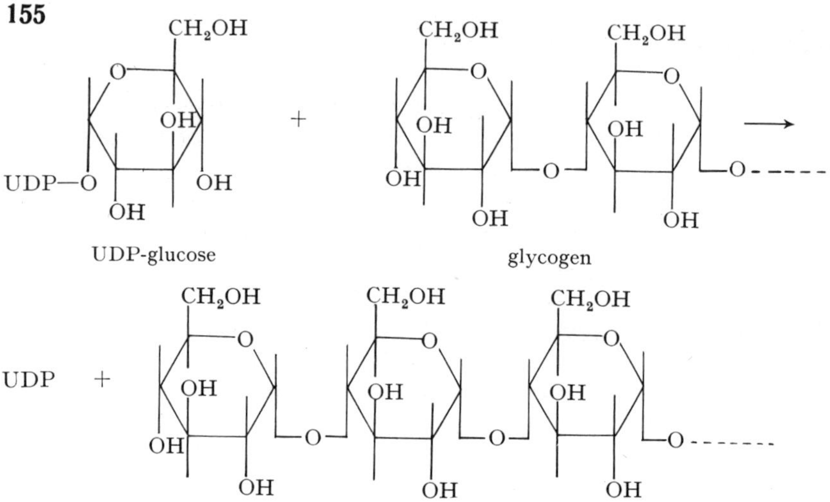

[UDP-glucose + glycogen → UDP + glycogen (extended)]

Glycogen is an amylopectin which is very highly branched. Other enzymes not considered here, called branching enzymes, produce the 1-6 links necessary for branch points.

Glycogen can be utilized to furnish glucose for the organism. In the liver, the breakdown of glycogen to glucose can take place when the glycoside bond between the glucose units is broken by the addition of phosphate. The nonreducing terminal glucosyl group is released as glucose-1-phosphate. This reaction is called phosphorolysis and is analogous to hydrolysis.

156

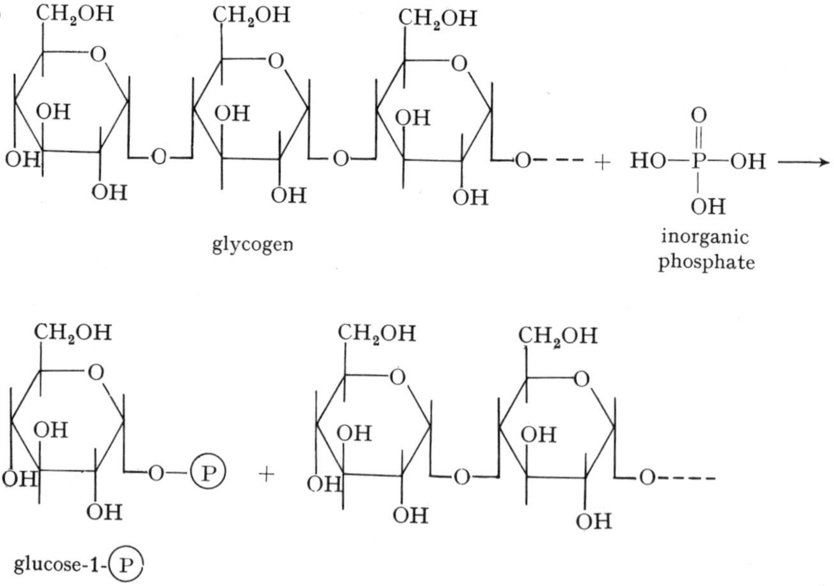

[glycogen + inorganic phosphate (HO—P(=O)(OH)—OH) → glucose-1-$\circled{P}$ + glycogen (shortened)]

Glucose-1-phosphate is converted to glucose-6-phosphate by the enzyme phosphoglucomutase (**153**). The hydrolysis of glucose-6-phosphate catalyzed by phosphatases within the liver yields glucose and inorganic phosphate.

157

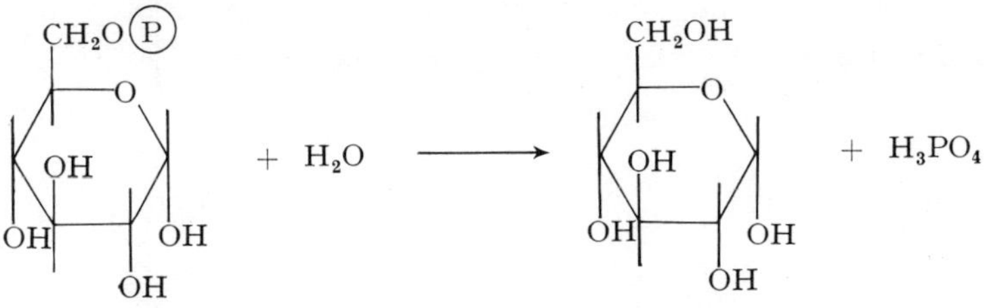

glucose-6-P + H_2O ⟶ glucose + H_3PO_4

This is *not* the reversal of **152**.

Except in the lactating female, galactose is metabolized almost exclusively by the liver. Galactose-1-phosphate is formed by reaction with adenosine triphosphate.

158

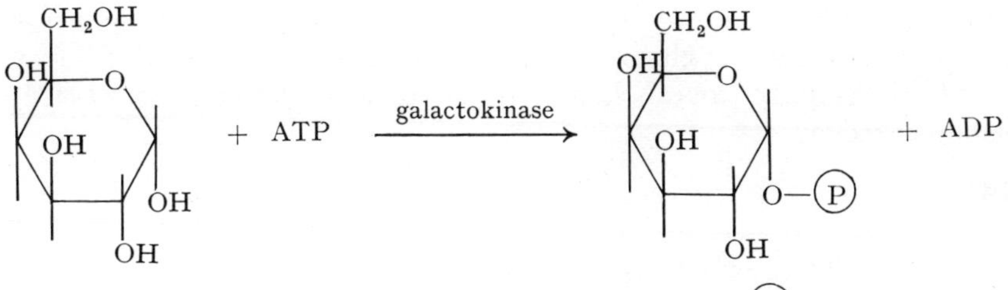

galactose + ATP $\xrightarrow{\text{galactokinase}}$ galactose-1-P + ADP

Galactose-1-phosphate enters an exchange reaction with uridine diphosphate glucose. Galactose-1-phosphate uridyl transferase (GPUT) catalyzes the reaction.

159

UDP-glucose + galactose-1-P $\xrightleftharpoons{\text{GPUT}}$ UDP-galactose + glucose-1-P

The galactosyl group is epimerized, probably though the formation of a ketone.

160

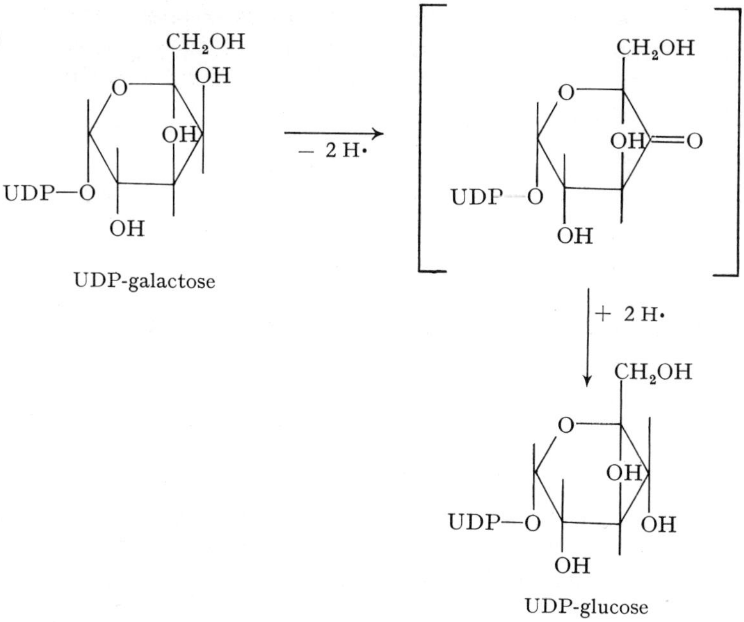

[UDP-galactose] $\xrightarrow{-2H\cdot}$ [oxidized intermediate] $\xrightarrow{+2H\cdot}$ [UDP-glucose]

The uridine diphosphate glucose can be used in the exchange reaction with galactose-1-phosphate (**159**), for glycogen synthesis (**155**), or released as glucose-1-phosphate (**156**).

Carbohydrate metabolism presented thus far can be briefly summarized as

161

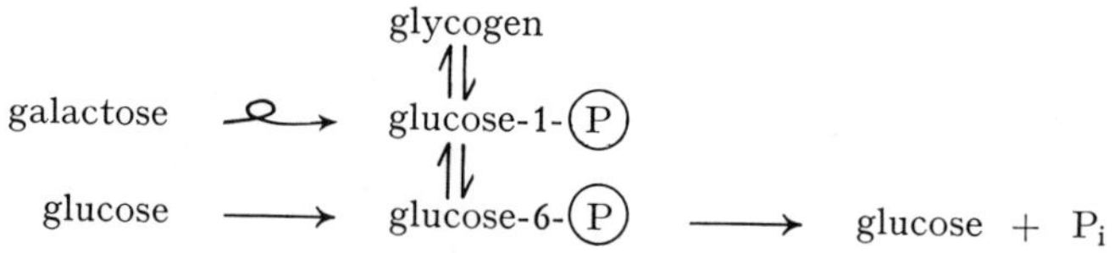

$$\text{galactose} \rightleftharpoons \text{glucose-1-}\textcircled{P} \rightleftharpoons \text{glycogen}$$
$$\text{glucose} \rightarrow \text{glucose-6-}\textcircled{P} \rightarrow \text{glucose} + P_i$$

To proceed, fructose-6-phosphate is formed by isomerization of glucose-6-phosphate.

162

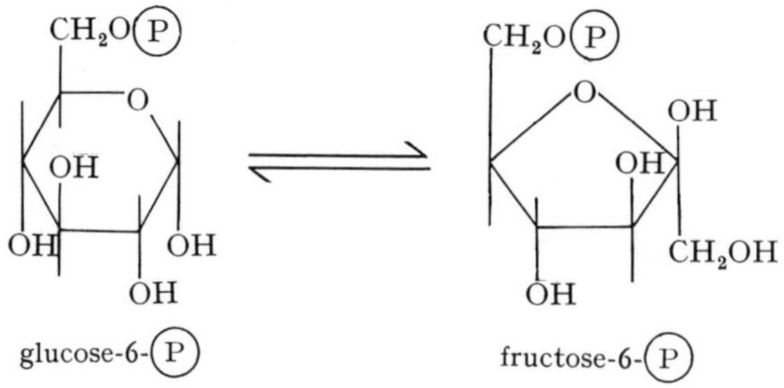

glucose-6-$\textcircled{P}$ $\rightleftharpoons$ fructose-6-$\textcircled{P}$

Dietary fructose is also converted to fructose-6-phosphate. This is accomplished by a transphosphorylation reaction with adenosine triphosphate.

163

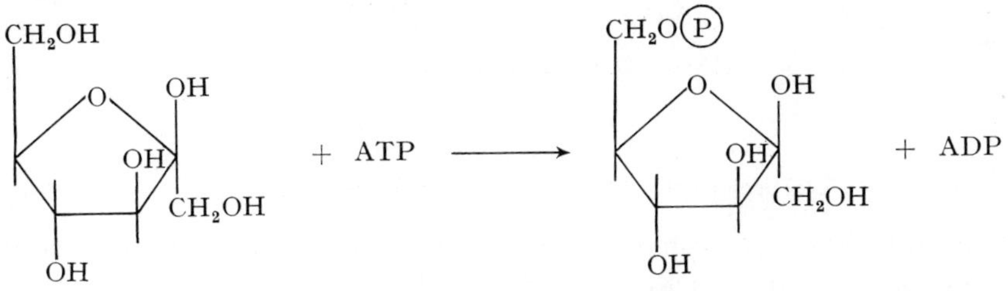

fructose + ATP ⟶ fructose-6-(P) + ADP

In the next step, fructose-6-phosphate is phosphorylated with adenosine triphosphate to produce fructose-1, 6-diphosphate.

164

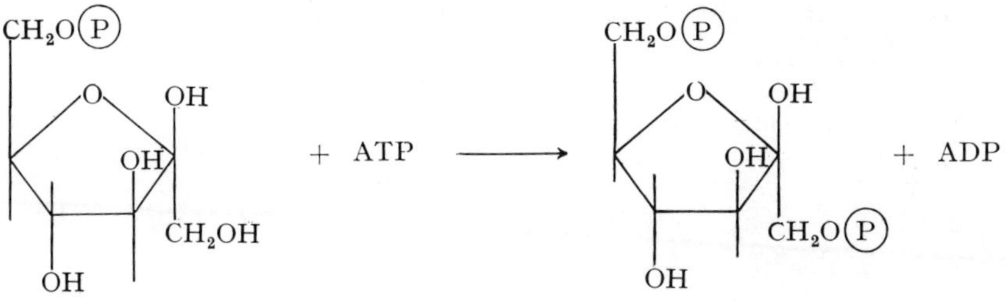

fructose-6-(P) + ATP ⟶ fructose-1,6-diphosphate + ADP

The enzyme aldolase severs the diphosphorylated fructose into two phosphorylated trioses, dihydroxyacetone phosphate and 3-phosphoglyceraldehyde, and another enzyme equilibrates the two triose phosphates.

165

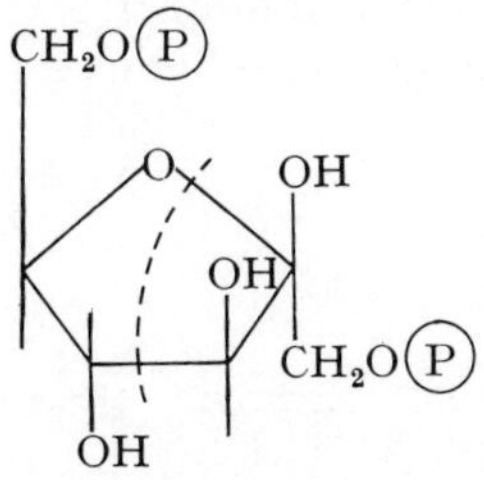

fructose-1,6-diphosphate

(Reaction continued on p. 104.)

$$\text{3-phosphoglyceraldehyde} \quad \begin{array}{c} \text{CHO} \\ | \\ \text{H—C—OH} \\ | \\ \text{CH}_2\text{O}\circled{P} \end{array} \rightleftharpoons \begin{array}{c} \text{CH}_2\text{OH} \\ | \\ \text{C=O} \\ | \\ \text{CH}_2\text{O}\circled{P} \end{array} \quad \text{dihydroxyacetone}\circled{P}$$

Glyceraldehyde phosphate is oxidized in a complex reaction and then forms a mixed acid anhydride. The same enzyme, represented as E—SH, catalyzes both reactions.

166

$$\begin{array}{c} \text{H} \quad \text{O} \\ \diagdown\!\!\diagup \\ \text{C} \\ | \\ \text{H—C—OH} \\ | \\ \text{CH}_2\text{O}\circled{P} \end{array} + \text{HS—E} \rightleftharpoons \left[\begin{array}{c} \text{H} \\ | \\ \text{HO—C—S—E} \\ | \\ \text{H—C—OH} \\ | \\ \text{CH}_2\text{O}\circled{P} \end{array}\right]$$

glyceraldehyde-3-phosphate

$$\begin{array}{c} \text{O} \quad \text{O} \\ \diagdown\!\!\!\parallel \quad \parallel \\ \text{C—O} \sim \text{P—OH} \\ | \quad\quad | \\ \text{H—C—OH} \quad \text{OH} \\ | \\ \text{CH}_2\text{O}\circled{P} \end{array} \xrightleftharpoons[\pm \text{HS—E}]{\pm \text{H}_3\text{PO}_4} \left[\begin{array}{c} \text{O=C—S—E} \\ | \\ \text{H—C—OH} \\ | \\ \text{CH}_2\text{O}\circled{P} \end{array}\right] \begin{array}{c} \longrightarrow \text{NAD}^+ \\ \longrightarrow \text{NAD·H + H}^+ \end{array}$$

1,3-diphosphate glyceraldehyde

Note the similarity in structure of 1,3-diphosphoglyceric acid and adenosine triphosphate. The anhydride structure of 1,3-diphosphoglyceric acid is a high-energy bond that can be transferred to adenosine diphosphate.

167

$$\begin{array}{c} \text{O} \\ \parallel \\ \text{C—O} \sim \circled{P} \\ | \\ \text{H—C—OH} \\ | \\ \text{CH}_2\text{O}\circled{P} \end{array} + \text{ADP} \rightleftharpoons \begin{array}{c} \text{O} \\ \parallel \\ \text{C—OH} \\ | \\ \text{H—C—OH} \\ | \\ \text{CH}_2\text{O}\circled{P} \end{array} + \text{ATP}$$

1,3-diphosphoglyceric acid   3-phosphoglyceric acid

The product, 3-phosphoglyceric acid, is isomerized to 2-phosphoglyceric acid, which is then dehydrated by the enzymatic action of enolase. Dehydration gives rise to 2-phosphoenolpyruvic acid, a structure containing another high-energy bond.

168

$$\begin{array}{c} COOH \\ | \\ H-C-OH \\ | \\ CH_2O-\textcircled{P} \end{array} \rightleftharpoons \begin{array}{c} COOH \\ | \\ H-C-O-\textcircled{P} \\ | \\ CH_2OH \end{array} \underset{\pm H_2O}{\overset{enolase}{\rightleftharpoons}} \begin{array}{c} COOH \quad O \\ | \qquad \quad \| \\ C-O \sim P-OH \\ \| \qquad \quad | \\ CH_2 \qquad OH \end{array}$$

3-phosphoglyceric acid   2-phosphoglyceric acid   2-phosphoenolpyruvic acid

The high-energy phosphate ester of 2-phosphoenolpyruvic acid is transferred to adenosine diphosphate, forming adenosine triphosphate and enolpyruvic acid. Tautomeric change of enolpyruvic acid yields the keto form, known as pyruvic acid.

169

$$\begin{array}{c} COOH \\ | \\ C-O\sim\textcircled{P} \\ \| \\ CH_2 \end{array} + ADP \rightleftharpoons \begin{array}{c} COOH \\ | \\ C-OH \\ \| \\ CH_2 \end{array} + ATP$$

enolpyruvic acid

$$\updownarrow$$

$$\begin{array}{c} O \\ \| \\ C-OH \\ | \\ C=O \\ | \\ CH_3 \end{array}$$

pyruvic acid

The final reaction in this scheme for carbohydrate metabolism is the reduction of pyruvic acid to lactic acid.

170

$$\begin{array}{c} COOH \\ | \\ C=O \\ | \\ CH_3 \end{array} + NAD \cdot H + H^+ \longrightarrow \begin{array}{c} COOH \\ | \\ H-C-OH \\ | \\ CH_3 \end{array} + NAD^+$$

pyruvic acid   lactic acid

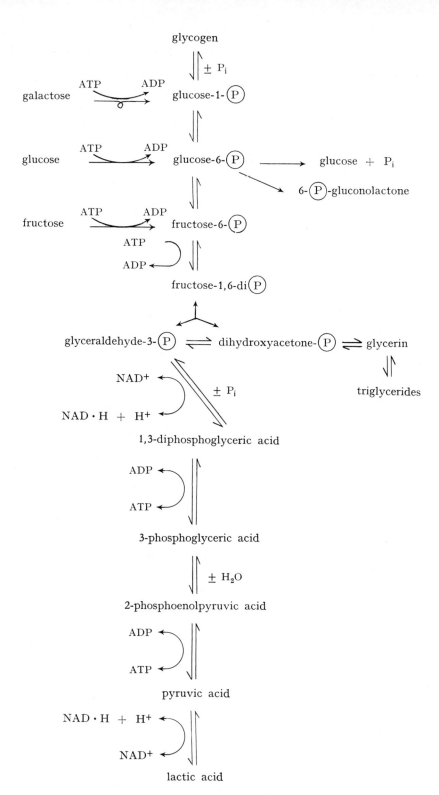

Fig. 7-1. Anaerobic glycolysis.

ROLE OF GLYCOLYSIS

The foregoing structures and reactions have been greatly simplified. Most of the names of the enzymes necessary for the individual reactions have been deleted. A simplified presentation of the scheme is complete enough to illustrate the importance of the system (see Fig. 7-1).

The system, called anaerobic glycolysis, provides a common metabolic pathway for the usual dietary carbohydrates. The system allows glucose, fructose, and galactose to be converted to glucose, glycogen, lactic acid, or any compound designated in the scheme. The reversibility of the scheme allows lactic acid to be converted to glycogen or glucose or both.

An interesting aspect of anaerobic glycolysis is that the components are almost invariably interconverted as phosphate derivatives. Another aspect to consider is that each hexose is oxidized to two moles of lactic acid, but no oxygen is utilized in the system.

An audit of the system reveals that two moles of adenosine triphosphate are required for entry of each hexose. This is true whether the starting point is glycogen or glucose. A mole of adenosine triphosphate is required to convert each mole of hexose into glycogen. When each mole of hexose is converted to two moles of lactate, there is a gain of four moles of adenosine triphosphate; two moles are gained in the conversion of 1,3-diphosphoglyceric acid to 3-phosphoenolpyruvic acid and two moles in the formation of pyruvic acid from 2-phosphoenolpyruvic acid. The actual net gain is two moles of adenosine triphosphate per mole of hexose, because there is an initial debit of two moles of adenosine triphosphate. The energy yield for anaerobic glycolysis is low, however, when compared with that obtained by the further oxidation of lactic acid.

One mole of NAD^+ is consumed in the oxidation of glyceraldehyde-3-phosphate, and one mole is produced in the reduction of pyruvic acid. Thus, an audit of electrons and protons yields a zero net balance.

OTHER PATHWAYS

Examination of Figure 7-1 reveals three branch points in the system. One is the hydrolysis of glucose-6-phosphate to glucose and inorganic phosphate. This pathway can be utilized to convert glycogen, galactose, and fructose to glucose. A second branch point is the oxidation of glucose-6-phosphate to 6-phosphogluconolactone. The continuation of this pathway will be presented later.

The third branch point of the scheme is the formation of glycerin from dihydroxyacetone. This branch provides a source of glycerin for triglyceride synthesis, and provides a connection between carbohydrate and lipid metabolism. The reduction of phosphodihydroxyacetone yields 3-phosphoglycerin, which upon hydrolysis by phosphatase releases glycerin. The glycerin phosphate and glycerin can both serve as sources of glycerin for triglyceride synthesis.

171

$$\begin{array}{c} CH_2OH \\ | \\ C=O \\ | \\ CH_2O\text{\textcircled{P}} \end{array} \underset{NAD^+}{\overset{NAD \cdot H + H^+}{\rightleftharpoons}} \begin{array}{c} CH_2OH \\ | \\ H-C-OH \\ | \\ CH_2O\text{\textcircled{P}} \end{array}$$

dihydroxyacetone phosphate   glycerin phosphate

$$\downarrow H_2O$$

$$\begin{array}{c} CH_2OH \\ | \\ H-C-OH \\ | \\ CH_2OH \end{array} + P_i$$

glycerin

METABOLIC CONTROLS

An important aspect of metabolism that can perhaps best be appreciated here is regulation. The enzyme-catalyzed reactions are fast, but metabolism does not occur with a bang, a ball of fire, and a cloud of smoke. Reference to Figure 7-1 reveals that some 13 enzymes and reactions are required for the conversion of hexoses to lactate. The small stepwise changes are advantageous to the organism because they permit control of the processes. The regulatory mechanisms are quite complex and will not be presented in detail, but some important aspects will be mentioned.

One regulatory mechanism is the inhibition of some enzymes by the product of the reaction. This type of inhibition was discussed in the chapter on enzymes.

Some enzymes that catalyze one reaction in a series of reactions can be inhibited by the product of another reaction of the series. For example, in **172,** product E exerts an inhibitory effect on the enzyme catalyzing the conversion of A to B.

172 $\quad A \xrightarrow{\text{enzyme}} B \longrightarrow C \longrightarrow D \longrightarrow E$

In this manner, an enzyme at the beginning of metabolic sequence can be regulated by a product produced later in the series. This type of regulation is called "negative feedback," a term used in the field of electronics. When adequate amounts of product accumulate, inhibition is increased and the system is slowed or stopped. A decrease in product concentration decreases inhibition and the rate of the system increases.

The availability of cofactors is another mechanism of system regulation. For example, in anaerobic glycolysis, a limited amount of NAD^+ would limit the rate at which 3-phosphoglyceraldehyde is converted to 1,3-diphosphoglyceric acid. Similar control can be effected by the availability of adenosine diphosphate and inorganic phosphate in this scheme. Hormones also exert great regulatory influence on metabolic processes.

Without controls, large quantities of hexose would produce correspondingly large amounts of lactic acid, glycogen, and glycerin if their rate of formation was solely dependent on the hexose concentration. With regulatory mechanisms, however, large amounts of glucose generally are converted to glycogen and stored until needed. If energy is required and the oxygen supply is limited, the glucose is converted to lactic acid with concomitant energy production.

Perhaps the best example of a lack of metabolic regulation is cancer. Malignant cells apparently have little growth regulation and may continue to accumulate until the host is destroyed, unless they are treated.

HEXOSE MONOPHOSPHATE SHUNT

In the earlier discussion of anaerobic glycolysis the formation of 6-phosphogluconolactone from the oxidation of glucose-6-phosphate was considered the second branch point (Fig. 7-2). From this point, the pathway leads to the pentoses and is called the hexose monophosphate shunt. Hydrolysis of the lactone yields the corresponding acid **(173)**.

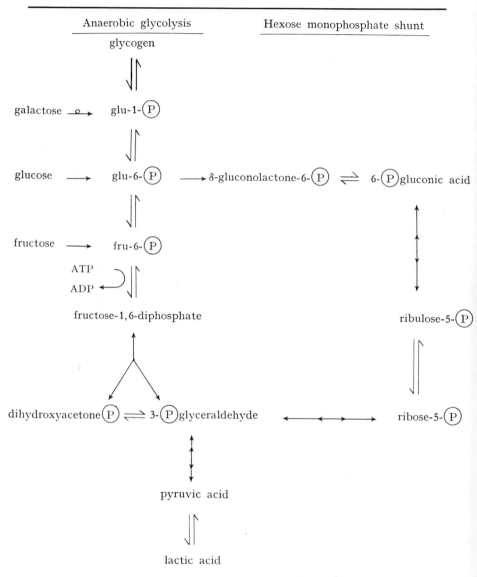

Fig. 7-2. Anaerobic glycolysis and hexose monophosphate shunt.

Oxidation and decarboxylation of the acid yields D-ribulose-5-phosphate. Epimerization of the ribulose yields D-ribose-5-phosphate, which is used for the biosynthesis of ribonucleosides and ribonucleotides. The corresponding deoxy derivatives are formed by reduction of the ribonucleosides and ribonucleotides.

173

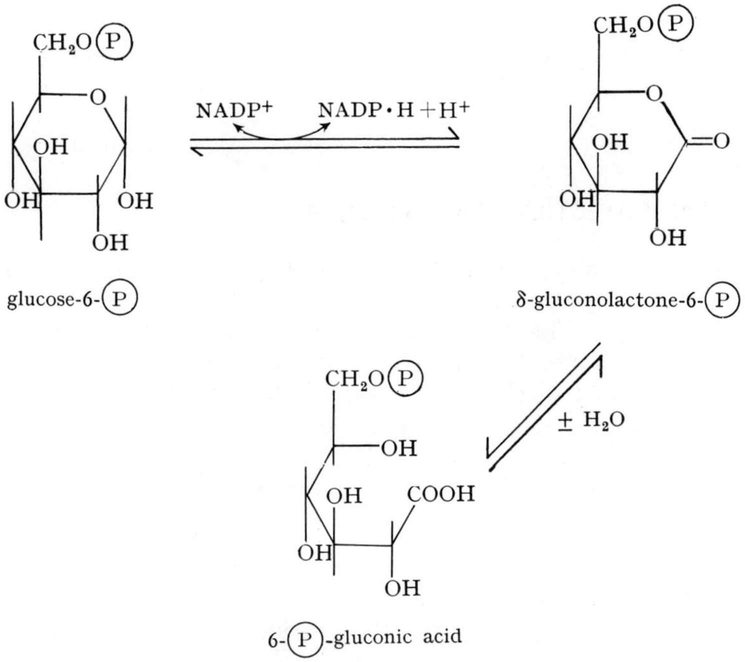

glucose-6-(P) ⇌ [NADP$^+$ / NADP·H + H$^+$] δ-gluconolactone-6-(P)

⇌ ± H$_2$O

6-(P)-gluconic acid

174

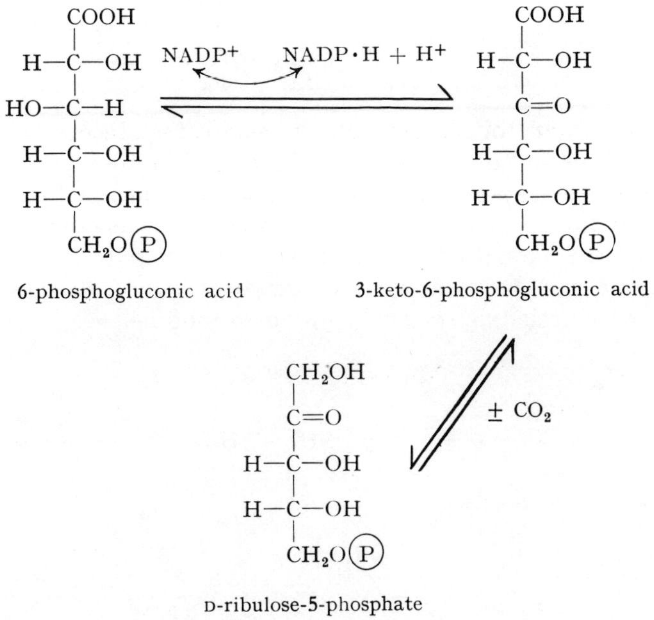

6-phosphogluconic acid ⇌ [NADP$^+$ / NADP·H + H$^+$] 3-keto-6-phosphogluconic acid

⇌ ± CO$_2$

D-ribulose-5-phosphate

Through a series of reactions, ribose-5-P can be converted to glyceraldehyde-3-P.

175   ribose-5-P $\longrightarrow$ $\longrightarrow$ $\longrightarrow$ glyceraldehyde-3-P

ENZYMATIC DETECTION OF GLUCOSE

Glucose dehydrogenase (glucose oxidase), an enzyme isolated from microorganisms, provides a method for measuring glucose in the clinical laboratory. The enzyme will catalyze the oxidation of glucose to gluconolactone, but the reaction requirements are different than those described in the monophosphate shunt. The enzyme is different because it reacts with glucose and requires a different coenzyme for activity. In the reaction, the protons and electrons are transferred to oxygen to produce hydrogen peroxide.

176   glucose + O_2 $\xrightarrow{\text{glucose dehydrogenase}}$ δ-gluconolactone + H_2O_2

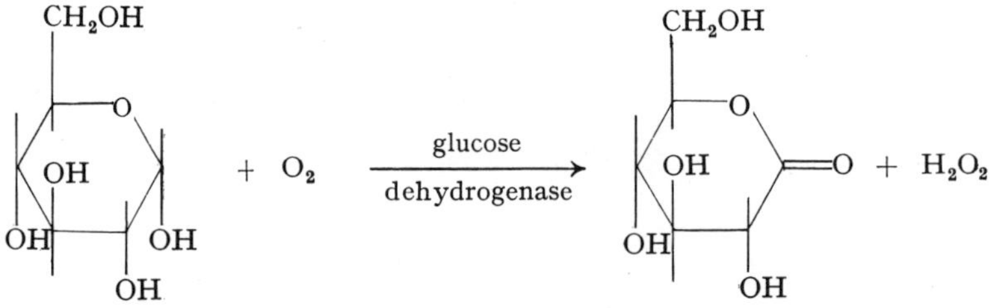

This specificity of glucose dehydrogenase for glucose is utilized in the detection and quantitation of glucose. The method for quantitation usually employs the measurement of the hydrogen peroxide. Peroxidase is commonly added to the reaction medium to catalyze the oxidation of a colorless compound to a colored derivative that can be measured colorimetrically. An example of this type of reaction is shown below for *o*-tolidine.

177   *o*-tolidine + H_2O_2 $\xrightarrow{\text{peroxidase}}$ colored product + $2 H_2O$

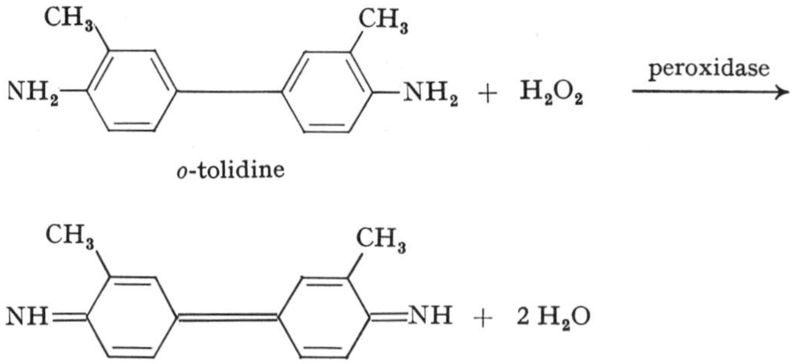

The enzymes, glucose dehydrogenase and peroxidase, can be used in typical " wet " chemical procedures or sorbed on a support medium, such as filter paper. The stability of the enzymes and chromogen permits the use of support media. The impregnated support media are known as " dip sticks " and are employed for the assay of glucose when great accuracy is not required. The dip stick is immersed into the solution to be assayed and the color developed compared to a standard color chart. This is a convenient method for estimating urine glucose.

GLYCOGEN STORAGE AND BLOOD GLUCOSE

The glycolytic scheme apparently occurs in most organisms. However, in mammals there are certain differences in various tissues. The liver is one of the main storage sites for glycogen. Following the ingestion of carbohydrates, the quantities of the various hexoses exceed the immediate need. The excess is converted by the liver to glycogen. Later, when the organism has a requirement for additional glucose, the glycogen can be converted to the phosphorylated glucoses and the hydrolysis of glucose-6-(P) yields glucose which then enters the bloodstream to be distributed to the various body cells.

Another storage site for glycogen is skeletal muscle. The metabolic scheme in muscle is very similar to that of the liver, with one notable exception: skeletal muscles do not have the phosphatase enzyme for hydrolyzing glucose-6-phosphate to glucose and inorganic phosphate. Therefore, muscle cannot supply glucose to the bloodstream and other tissues directly.

There are several terms applied to the metabolism of carbohydrates with which the student should become familiar. Glycolysis refers to the breakdown of carbohydrates to lactic acid. Glycogenolysis refers to the breakdown of glycogen to yield glucose, and glycogenesis indicates glycogen synthesis.

ENZYME DEFICIENCIES

An inherited deficiency of certain enzymes can produce benign or profound effects on organisms. Deficiencies of this type are sometimes referred to as " inborn errors of metabolism."

In the condition called galactosemia, the enzyme galactose-1-(P)-uridyl transferase (GPUT) **(159)** is absent or insufficient. The enzymatic conversion of galactose to glucose is blocked and the concentration of blood galactose elevated. High concentrations of blood galactose are toxic, and infants with hypergalactosemia have diarrhea, gastric distress, and irritability. Prolonged hypergalactosemia can result in irreversible damage to the central nervous system. It is therefore imperative that the activity of this enzyme be measured in all newborns. The treatment for galactosemia is removal of the source of dietary galactose. The major source is milk so a nonmilk diet is used.

Individuals with galactosemia develop an alternate metabolic pathway as they mature.

178
$$\text{UTP} + \text{gal-1-}\textcircled{P} \xrightleftharpoons{\text{UDGP}} \text{UDP-gal} + \text{P–P}_i$$
pyrophosphate

At birth uridine diphosphogalactose pyrophosphorylase (UDGP) activity is low, but later in life there is enough to metabolize ordinary amounts of galactose and the intolerance disappears.

Normally, galactose-1-phosphate uridyl transferase and other enzymes for assay are present in erythrocytes. To evaluate GPUT, hemolyzed blood is added to a solution containing galactose-1-phosphate, uridine diphosphoglucose, and NADP$^+$. If enzymatic activity is present, NADP · H is formed and can be measured or detected or both fluorometrically. Qualitative detection of the enzyme using the reactions shown below can be done on filter paper.

179
$$\text{gal-1-}\textcircled{P} + \text{UDP-glu} \rightleftharpoons \text{glu-1-}\textcircled{P} + \text{UDP-gal}$$

$$\text{glu-1-}\textcircled{P} \rightleftharpoons \text{glu-6-}\textcircled{P}$$

$$\text{glu-6-}\textcircled{P} \xrightarrow[\text{glucose dehydrogenase}]{\text{NADP}^+ \quad \text{NADP·H} + \text{H}^+} \delta\text{-gluconolactone-6-}\textcircled{P}$$

Individuals who excrete large amounts of L-xylose in their urine have a hereditary enzyme deficiency called pentosuria. It is apparently an innocuous condition. Xylose, a ketopentose, is formed from D-ribulose-5-phosphate, but the enzyme required for further metabolism is absent.

CHAPTER 8

LIPID METABOLISM

A major portion of mammalian fat is stored in fat depots in the form of triglycerides. Triglycerides can be synthesized from fatty acids and glycerin or hydrolyzed to glycerin and fatty acids.

180

$$\begin{array}{c} CH_2O-\overset{O}{\overset{\|}{C}}-R \\ | \\ CHO-\overset{O}{\overset{\|}{C}}-R \\ | \\ CH_2O-\overset{O}{\overset{\|}{C}}-R \end{array} \quad \underset{\rightleftharpoons}{\pm 3 H_2O} \quad \begin{array}{c} CH_2OH \\ | \\ CHOH \\ | \\ CH_2OH \end{array} + 3 R-COOH$$

triglycerides glycerin fatty acids

In the previous chapter, reference was made to the relationship of glycerin to carbohydrate metabolism. Here, the metabolism of free fatty acids will be discussed.

FATTY ACID METABOLISM

The first step in fatty acid metabolism is esterification with coenzyme A, represented as CoASH.

181 $CH_3-(CH_2)_x-CH_2-CH_2-COOH + CoASH + ATP \longrightarrow$

$$CH_3-(CH_2)_x-CH_2-CH_2-\overset{O}{\overset{\|}{C}}-S-CoA + AMP + P-P_i$$

CoA is a complex nucleotide derivative containing adenosine diphosphate and the vitamin pantothenic acid. It functions as a coenzyme for most fatty acid and many of the dicarboxylic acid reactions. The substrate molecules become reactive when they form thioesters with CoA. Coenzyme A does not undergo a valence change as does $NADP^+$ and NAD^+. Its functions are analogous to the phosphates in carbohydrate metabolism.

The fatty acid coenzyme A complexes are first oxidized to yield the α,β-unsaturated derivatives. The coenzyme required here is known as FAD (flavin adenine dinucleotide) which contains another vitamin, riboflavin. Hydration of the β-unsaturated coenzyme A derivative yields the corresponding β-hydroxy compound.

182

$$CH_3-(CH_2)_x-CH_2-CH_2-CH_2-\underset{\underset{O}{\|}}{C}-SCoA + FAD \rightleftharpoons$$

$$CH_3-(CH_2)_x-CH_2-CH=CH-\underset{\underset{O}{\|}}{C}-SCoA + FAD \cdot H_2$$

$$\Big\updownarrow + H_2O$$

$$CH_3-(CH_2)_x-CH_2-\underset{OH}{\underset{|}{CH}}-CH_2-\underset{\underset{O}{\|}}{C}-SCoA$$

Next in the sequence is the oxidation of the β-hydroxy compound to yield the β-keto compound. This reaction requires the coenzyme NAD^+.

183

$$CH_3-(CH_2)_x-CH_2-\underset{OH}{\underset{|}{CH}}-CH_2-\underset{\underset{O}{\|}}{C}-SCoA + NAD^+ \rightleftharpoons$$

$$CH_3-(CH_2)_x-CH_2-\underset{\underset{O}{\|}}{\overset{\overset{O}{\|}}{C}}-CH_2-\underset{\underset{O}{\|}}{C}-SCoA + NAD \cdot H + H^+$$

The reaction series is complete with cleavage of the β-keto compound with coenzyme A to produce acetyl coenzyme A and the new fatty acid derivative which differs from the original by two carbons.

184

$$CH_3-(CH_2)_x-CH_2-\underset{\underset{O}{\|}}{C}-CH_2-\overset{O}{\underset{\|}{C}}-SCoA + CoASH \rightleftharpoons$$

$$CH_3-(CH_2)_{x-2}-CH_2-\overset{O}{\underset{\|}{C}}-SCoA + CH_3-\overset{O}{\underset{\|}{C}}-SCoA$$

The new shorter-chain fatty acid coenzyme A compound can reenter the reaction scheme (**182**) to yield an additional mole of acetyl coenzyme A, and the chain length is again reduced by two carbons.

FATTY ACID SYNTHESIS

The metabolism of fatty acids summarized in Figure 8-1 serves as the degradative scheme for the entire cell, but as a synthetic

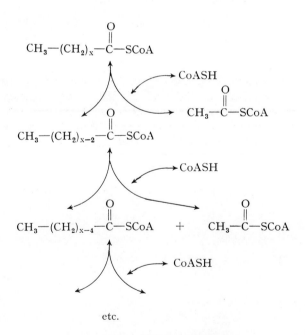

Fig. 8-1. Fatty acids metabolism.

pathway only in mitochondria. In cytoplasm, the biosynthesis of fatty acids begins with the formation of a malonic acid derivative. This reaction is an example of CO_2 fixation.

185

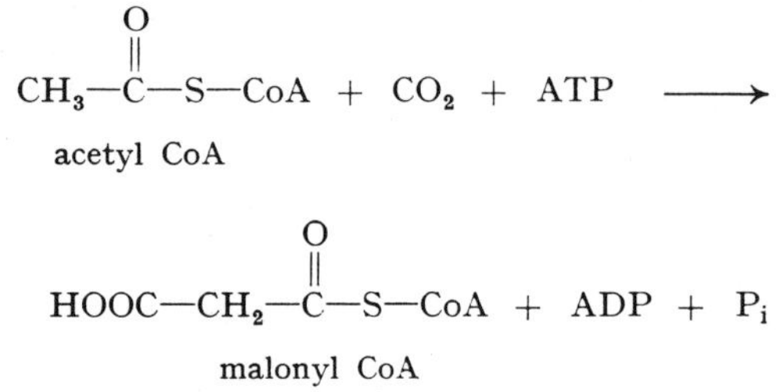

$$CH_3-\overset{O}{\underset{\|}{C}}-S-CoA + CO_2 + ATP \longrightarrow$$
acetyl CoA

$$HOOC-CH_2-\overset{O}{\underset{\|}{C}}-S-CoA + ADP + P_i$$
malonyl CoA

Condensation of malonyl CoA and acetyl CoA, with decarboxylation yields acetoacetyl CoA.

186

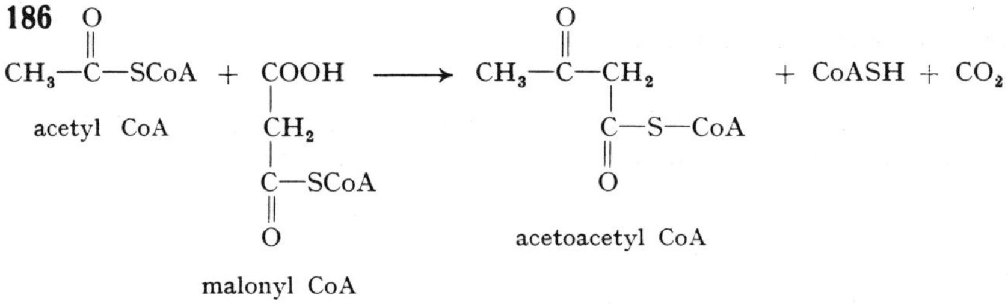

$$CH_3-\overset{O}{\underset{\|}{C}}-SCoA + \underset{\underset{O}{\|}}{\underset{C-SCoA}{\underset{|}{\underset{CH_2}{\underset{|}{COOH}}}}} \longrightarrow CH_3-\overset{O}{\underset{\|}{C}}-CH_2 \atop \underset{O}{\underset{\|}{C-S-CoA}} + CoASH + CO_2$$

acetyl CoA   malonyl CoA   acetoacetyl CoA

Butyryl CoA is formed by the reduction of acetoacetyl CoA via the reaction series outlined in **187** through **189**. These reactions are the same as those above for the formation of fatty acids from acetyl CoA, but the coenzyme requirements are different. The condensation of butyryl CoA with malonyl CoA (**190**) is the same reaction as for acetoacetyl CoA formation (**186**).

187 $CH_3-\underset{\underset{O}{\|}}{C}-CH_2COSCoA + NADP \cdot H + H^+ \rightleftharpoons$

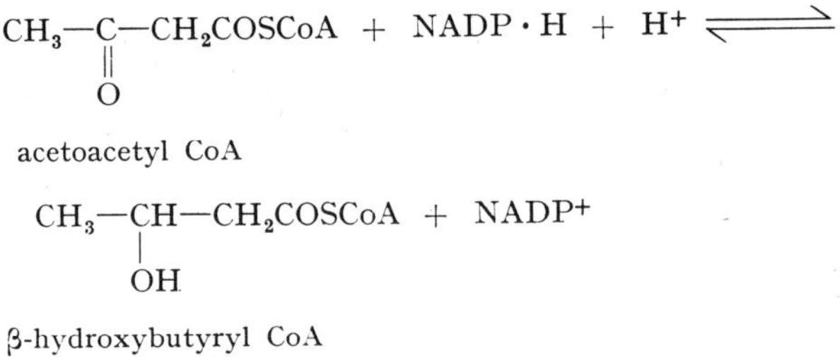

acetoacetyl CoA

$CH_3-\underset{\underset{OH}{|}}{CH}-CH_2COSCoA + NADP^+$

β-hydroxybutyryl CoA

188 $CH_3-\underset{\underset{OH}{|}}{CH}-CH_2COSCoA \xrightleftharpoons{\pm H_2O} CH_3-CH=CH-COSCoA$

β-hydroxybutyryl CoA   crotonyl CoA

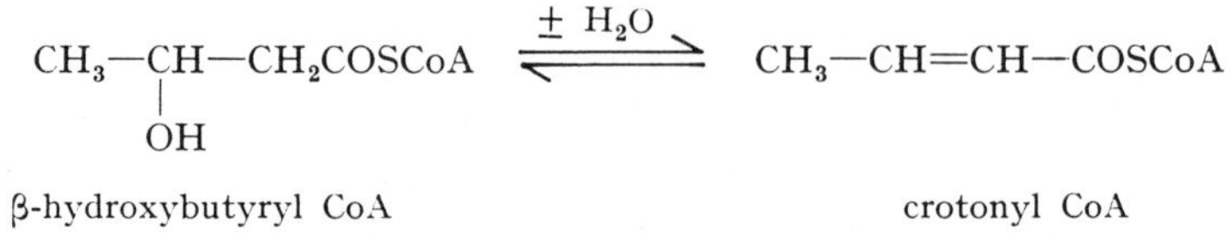

189 $CH_3CH=CH-COSCoA + NADP \cdot H + H^+ \rightleftharpoons$
crotonyl CoA

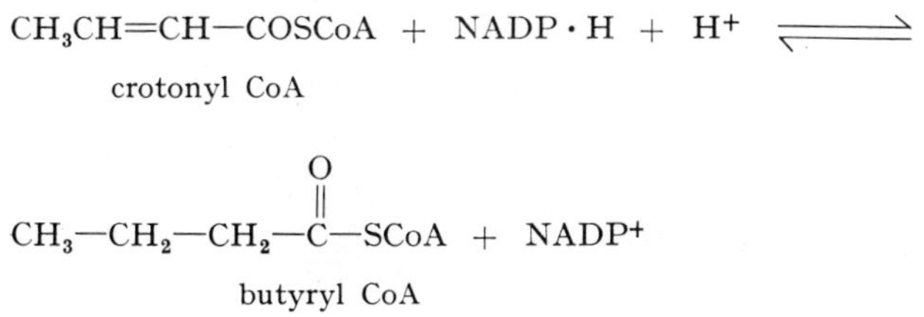

$$CH_3-CH_2-CH_2-\overset{O}{\underset{\|}{C}}-SCoA + NADP^+$$
butyryl CoA

190 $CH_3-CH_2-CH_2-\overset{O}{\underset{\|}{C}}-SCoA +\ \begin{array}{c}COOH\\|\\CH_2\\|\\C-S-CoA\\\|\\O\end{array} \rightleftharpoons$

butyryl CoA   malonyl CoA

$$CH_3-CH_2-CH_2-\overset{O}{\underset{\|}{C}}-\begin{array}{c}CH_2\\|\\C-S-CoA\\\|\\O\end{array} + CO_2 + CoASH$$

β-ketohexanoyl CoA

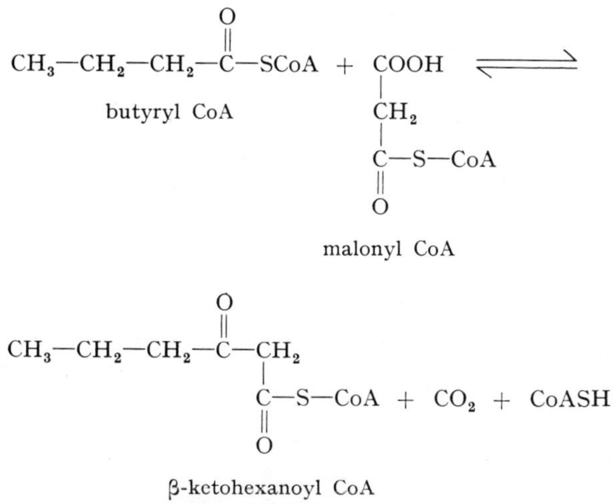

β-Ketohexanoyl CoA is reduced as before, followed by further additions and reductions to yield the long-chain fatty acid CoA compounds. These derivatives are used in triglyceride synthesis.

GLYCERIDE SYNTHESIS

Glycerin phosphate, a product of anaerobic glycolysis, reacts with the fatty acid acyl CoA to form phosphatidic acid **(191)**. Then phosphatidic acid can be converted to phospholipids or to triglycerides by the action of phosphatase and reaction with fatty acid acyl CoA (see Fig. 8-2).

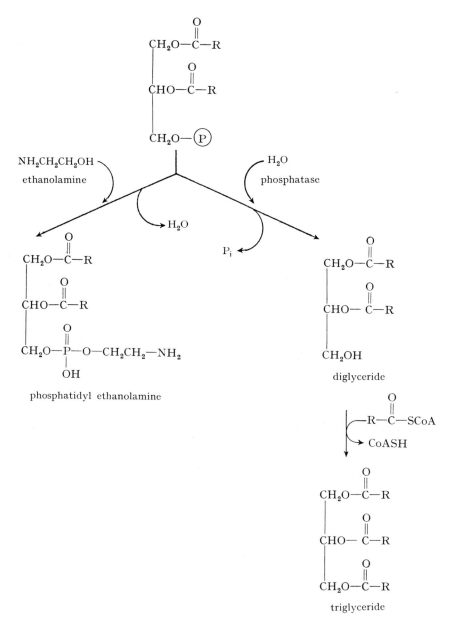

Fig. 8-2. Phospholipid and triglyceride synthesis.

191

$$2\ R-\overset{O}{\underset{\|}{C}}-S-CoA\ +\ \begin{array}{c}CH_2OH\\|\\CHOH\\|\\CH_2O\text{\textcircled{P}}\end{array}\ \longrightarrow\ \begin{array}{c}CHO-\overset{O}{\underset{\|}{C}}-R\\|\\CHO-\overset{O}{\underset{\|}{C}}-R\\|\\CH_2O\text{\textcircled{P}}\end{array}\ +\ 2\ CoASH$$

glycerin phosphate   phosphatidic acid

STEROL SYNTHESIS

Acetyl CoA and acetoacetyl CoA are intermediates in the biosynthesis of several compounds. One is the steroid cholesterol which is formed by the polymerization of acetyl CoA via acetoacetyl CoA, through several steps.

192

$$2\ CH_3-\overset{O}{\underset{\|}{C}}-SCoA\ \longrightarrow\ CH_3-\overset{O}{\underset{\|}{C}}-CH_2-\overset{O}{\underset{\|}{C}}-SCoA\ +\ CoASH$$

acetyl CoA   acetoacetyl CoA $\downarrow$

$\downarrow$

cholesterol

Cholesterol synthesis is regulated by a negative feedback system. If the diet contains large amounts of cholesterol then the organism will synthesize little. If, however, the diet contains only a small amount of cholesterol, the organism can produce considerable quantities. The amount synthesized is in part directly related to cholesterol content of the diet.

Cholic acid, vitamin D, and steroid hormones are products of cholesterol. Cholic acid and vitamin D have been discussed previously, and the steroid hormones will be considered in the chapter on hormones.

KETOGENESIS

The liver is the primary organ concerned with fat metabolism. One of its major functions is to convert fatty acids to acetoacetate and β-hydroxybutyrate for utilization by extrahepatic tissues. The process is termed ketogenesis.

Acetoacetic acid and β-hydroxybutyric acid are products of liver ketogenesis for utilization by extrahepatic tissues.

193

$$2\ CH_3-\underset{O}{\overset{\|}{C}}-SCoA \rightleftarrows CH_3-\underset{O}{\overset{\|}{C}}-CH_2-\underset{O}{\overset{\|}{C}}-S-CoA$$
acetoacetyl CoA

$$\xrightarrow{H_2O}\ CH_3-\underset{O}{\overset{\|}{C}}-CH_2-\underset{O}{\overset{\|}{C}}-OH\ +\ CoASH$$
acetoacetic acid

$$\underset{NAD^+}{\overset{NAD\cdot H\ +\ H^+}{\rightleftarrows}}$$

$$CH_3-\underset{OH}{\overset{|}{C}H}-CH_2-COOH$$
β-hydroxybutyric acid

These two acids and acetone, formed by the spontaneous decarboxylation of acetoacetic acid, are known collectively as the ketone bodies. When the oxidation of acetyl CoA is impaired, its carbons are directed to the formation of the ketone bodies. The accumulation of these substances in blood in abnormal amounts is called ketonemia. When abnormally large amounts are excreted in the urine, it is termed ketonuria. The term ketosis, as applied to ketone body metabolism, is so nonspecific that for practical purposes it is meaningless.

In addition to the production of excess ketone bodies, acetyl CoA is also shunted to cholesterol synthesis. Hypercholesterolemia is a common finding in individuals with diabetes mellitus.

STRUCTURAL LIPIDS

Mammals employ lipids for structural purposes and as food reserves. Cell walls contain the structural lipids which are important in selective permeability. The brain and large nerve tracts have considerable amounts of lipid material. Some occur as glycolipids. An ascribed function of the nerve system lipid is insulation, both mechanical and electrical.

FAT STORAGE

The deposition of triglycerides as adipose tissue is a well known phenomenon. When ample food is available, most humans are prone to consume quantities in excess of their needs. It will be shown later that carbohydrates and proteins can be converted to fat — a truism probably known to all from personal experience.

Acetoacetic acid cannot be utilized by the liver because the enzyme required for the conversion to acetoacetyl CoA is absent, and it cannot be metabolized unless it is esterified with CoA. Most other cells, however, readily utilize acetoacetic acid either for synthetic purposes or for energy. Some energy is produced from the degradation of fatty acids to acetyl CoA, but the major portion is derived from further oxidation.

Additional oxidation of acetyl CoA requires concomitant carbohydrate metabolism. If carbohydrate metabolism is impaired, fat metabolism is accelerated, apparently as a compensatory mechanism. The condition occurs in starvation, malnutrition, and diabetes mellitus. The chapter on common metabolic pathways contains a more thorough discussion of these conditions of acetyl CoA oxidation.

FATTY LIVER

Increased ketogenesis, including that due to hepatocellular toxins (chloroform, certain drugs), results in the increased mobilization of fat from the depots to the liver. If the rate of entrance of the fats into the liver exceeds the rate of utilization, the fats accumulate and lead to a condition referred to as fatty liver. The lipid content of a normal human liver is approximately 4%, but in fatty livers it may reach 30%. The accumulation of large quantities is not toxic but interferes mechanically with liver functions.

LIPID TRANSPORT

The vascular system is unable to transport lipids as solutes because they are water insoluble. The fats are transported from the intestine to the sites of storage or metabolism in combination with protein. The protein provides sufficient solubility so that the fats can be transported by the vascular system. The opalescent appearance of blood serum shortly after lipid ingestion (postabsorption lipemia) is due to the large concentration of these protein fat complexes. These small particles, about 1 μ in diameter, are called chylomicrons. Lipoproteins are normal constituents of blood and provide the mechanism for lipid and fatty acid transport between tissues.

BLOOD LIPIDS

In Table 8-1 the lipid fractions commonly measured in clinical chemistry laboratories are listed along with the average concentration range for normal individuals (" normal values ") in the fasting state.

TABLE 8-1. BLOOD LIPIDS

Fraction	Range
Total lipids	450–1000 mg/dl
Cholesterol, total	100–280 mg/dl
Triglycerides	30–135 mg/dl
Phospholipids	5–12 mg P_i/dl
Fatty acids*	450–900 μM

* Fatty acids are also called free fatty acids (FFA), nonesterified fatty acids (NEFA), and unesterified fatty acids (UFA).

In conditions where there is an elevation of one fraction, there is usually an elevation of the other fractions. There are individuals who demonstrate increased cholesterol and triglyceride values that are not known to be disease-related. These disorders are usually called hereditary hyperlipoproteinemias. They can be detected and separated into five distinct types by electrophoresis. The technique for electrophoretic separation is the same as for proteins except the protein fractions are not measured. Instead, the medium is subjected to a lipid dye, and lipid material can then be visualized and quantitated. Detections and identification of hyperlipemia by type is important, because some types can be treated and the treatment is type-specific.

CHAPTER 9

AMINO ACID METABOLISM

The amino acids from dietary proteins are absorbed and enter what is commonly referred to as the amino acid pool. This pool represents all the amino acids in the vascular system and other fluids which are available for utilization by the various tissues. Amino acids from this pool are used in the synthesis of proteins and nonprotein constituents and can be oxidized to produce CO_2, water, ammonia, and energy. When proteins are degraded, the amino acids reenter the pool. Some of the amino acids can be synthesized from carbohydrate metabolites.

194

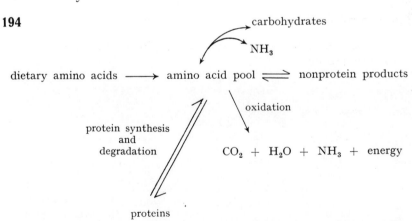

In carbohydrate metabolism, the usual dietary sugars all enter a common metabolic pathway, and the fatty acids are all degraded by a single process. However, each amino acid has its own more or less unique metabolic pathway. Only a few metabolic reactions common to most of the amino acids will be considered.

COMMON METABOLIC REACTIONS

OXIDATIVE DEAMINATION

Oxidative deamination is one reaction which is characteristic of the vast majority of amino acids.

195
$$R-\underset{NH_2}{\underset{|}{CH}}-COOH \xrightarrow[\text{(amino acid oxidase)}]{\text{amino acid dehydrogenase}} R-\underset{NH}{\underset{\|}{C}}-COOH + 2H\cdot$$

$$\downarrow H_2O$$

$$R-\underset{O}{\underset{\|}{C}}-COOH + NH_3$$

In this reaction, amino acids are oxidized by the removal of two protons and two electrons to yield the corresponding imino acid. The imino acid is then spontaneously hydrolyzed to the corresponding α-keto acid and ammonia.

TRANSAMINATION

Transamination is another type of reaction which is characteristic of most amino acids.

196
$$\begin{array}{c} COOH \\ | \\ H-C-NH_2 \\ | \\ CH_2 \\ | \\ COOH \end{array} + \begin{array}{c} COOH \\ | \\ C=O \\ | \\ CH_2 \\ | \\ CH_2 \\ | \\ COOH \end{array} \rightleftharpoons \begin{array}{c} COOH \\ | \\ H-C-NH_2 \\ | \\ CH_2 \\ | \\ CH_2 \\ | \\ COOH \end{array} + \begin{array}{c} COOH \\ | \\ C=O \\ | \\ CH_2 \\ | \\ COOH \end{array}$$

aspartic acid ・ α-ketoglutaric acid ・ glutamic acid ・ oxalacetic acid

In this particular enzymatic reaction, aspartic acid transfers its amino group to α-ketoglutaric acid and in turn receives a keto group to yield glutamic acid and oxalacetic acid as products. This particular reaction is clinically significant since it is used for diagnostic purposes, and it will be considered in the chapter on clinically important enzymes.

UREA SYNTHESIS

Ammonia produced from the catabolism of amino acids is toxic to mammals. A mechanism for its removal is the formation of urea (see Fig. 9-1). In this very simplified scheme, ornithine, a diamino acid, combines with CO_2 and ammonia to produce citrulline. Citrulline in turn condenses with ammonia to produce arginine. The latter, upon hydrolysis, yields urea in the enol form and ornithine. The

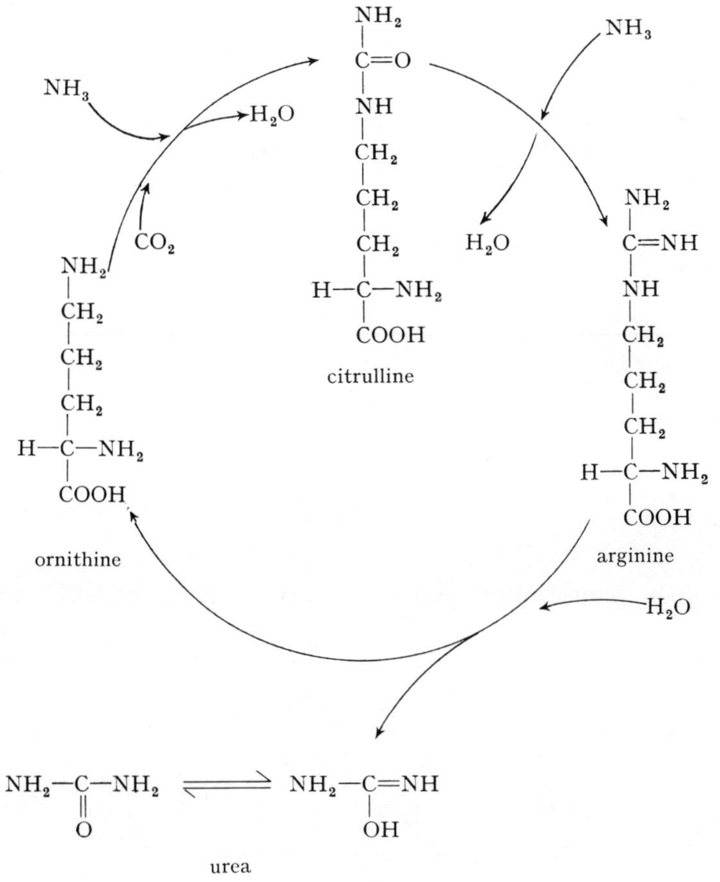

Fig. 9-1. Urea synthesis.

synthesis of urea requires energy and is another CO_2 fixing process. The synthesis of urea as described is confined to the liver.

CREATINE METABOLISM

Creatine is an important nonprotein product derived from certain amino acids. The initial step in the synthesis is a reaction between glycine and arginine. The amidine group from arginine is transferred to glycine to produce guanidinoacetic acid. This type of reaction is called transamidination – not to be confused with transamination.

197

$$\underset{\text{arginine}}{\begin{array}{c} NH_2 \\ | \\ C=NH \\ | \\ NH \\ | \\ (CH_2)_3 \\ | \\ HCNH_2 \\ | \\ COOH \end{array}} + \underset{\text{glycine}}{NH_2CH_2COOH} \xrightarrow{\text{transamidination}} \underset{\text{ornithine}}{\begin{array}{c} NH_2 \\ | \\ (CH_2)_3 \\ | \\ HCNH_2 \\ | \\ COOH \end{array}} + \underset{\text{guanidinoacetic acid}}{\begin{array}{c} NH_2 \\ | \\ C=NH \\ | \\ NH-CH_2COOH \end{array}}$$

Guanidinoacetic acid reacts with methionine in a transmethylation reaction to yield homocysteine and creatine.

198

$$\underset{\text{guanidinoacetic acid}}{\begin{array}{c} NH_2 \\ | \\ C=NH \\ | \\ HN-CH_2COOH \end{array}} + \underset{\text{methionine}}{\begin{array}{c} CH_3 \\ | \\ S \\ | \\ CH_2 \\ | \\ CH_2 \\ | \\ HCNH_2 \\ | \\ COOH \end{array}} \xrightarrow{\text{transmethylation}}$$

$$\underset{\text{homocysteine}}{\begin{array}{c} SH \\ | \\ CH_2 \\ | \\ CH_2 \\ | \\ HCNH_2 \\ | \\ COOH \end{array}} + \underset{\text{creatine}}{\begin{array}{c} NH_2 \\ | \\ C=NH \\ | \\ CH_3-N-CH_2COOH \end{array}}$$

Creatine performs a very important role of energy storage in muscles.

199

$$\begin{array}{c} NH_2 \\ | \\ C=NH \\ | \\ CH_3-N-CH_2COOH \end{array}$$
creatine

$$\begin{array}{c} O \\ \parallel \\ HN \sim P-OH \\ | \hspace{1em} \backslash \\ C=NH \hspace{0.5em} OH \\ | \\ CH_3-N-CH_2COOH \end{array}$$
phosphocreatine
(creatine phosphate)

Cycle: ATP → energy + P_i → ADP → ATP generating system + P_i → ATP

Creatine can react with ATP, a very reactive and unstable compound, to form the phosphoamide, phosphocreatine, or creatine phosphate. When this reaction occurs, the high energy stored in ATP is transferred to phosphocreatine, which in turn can be stored in muscles for long periods without appreciable decomposition.

When muscle is at rest, ATP is generated from oxidative phosphorylation or the other ATP-producing mechanisms, and it in turn reacts with creatine to form phosphocreatine. At times when muscle is doing work, particularly strenuous work, and the production of ATP cannot meet the energy demands of the muscle, ADP (which is produced in muscle) can react with phosphocreatine to produce ATP and creatine. Thus, phosphocreatine serves as a pool of available energy in that it can regenerate ATP from ADP.

The end product of creatine metabolism is creatinine, which is formed from creatine by dehydration.

200 (A)

$$\begin{array}{c} NH_2 \\ | \\ C=NH \\ | \\ CH_3-N-CH_2-COOH \end{array} \xrightarrow{-H_2O} \begin{array}{c} H-N- \\ | \hspace{2em} | \\ C=NH \hspace{1em} | \\ | \hspace{2em} | \\ CH_3-N-CH_2-C=O \end{array}$$

creatine $\hspace{8em}$ creatinine

(B)

$$\begin{array}{c} \hspace{2em} O \\ \hspace{2em} \parallel \\ CH_3-N \hspace{1em} NH \\ \hspace{2em} \parallel \\ \hspace{2em} NH \end{array}$$

creatinine

(*Note*: **200** (A) *is shown for clarity*, **200** (B) *is shown for preferred configuration.*)

There is no known enzyme that catalyzes the conversion of creatine to creatinine; it apparently is a spontaneous reaction.

The amount of creatinine excreted in the urine per day for a given individual on a controlled diet is quite constant, and this amount is directly proportional to the muscle mass of the individual. Adult males secrete from 1.2 to 2.0 g of creatinine per day, while females secrete from 1.0 to 1.8 g per day. Except in pregnancy and certain muscle diseases, creatine is normally found in only small concentrations in the urine of adults. However, in the urine of children, creatine is normally found in greater concentrations than creatinine until puberty, when it decreases.

Analysis of creatinine in urine can often be used to determine whether a urine specimen is a 24-hour collection. For instance, suppose a 24-hour collection from an adult male is found to contain 0.5 g of creatinine; this would be reason to suspect that the sample did not represent a true 24-hour collection.

METABOLISM OF PHENYLALANINE AND TYROSINE

The aromatic amino acids have many interesting metabolic reactions; a few of these in abbreviated form will be presented here.

A portion of the metabolic pathway for phenylalanine and tyrosine is represented in **201**.

In normal individuals with phenylalanine hydroxylase activity, tyrosine is formed and can be further metabolized to melanins or skin pigments. The lack of tyrosinase activity (B) produces a condition called albinism. Individuals with this condition (albinos) apparently suffer no physical distress.

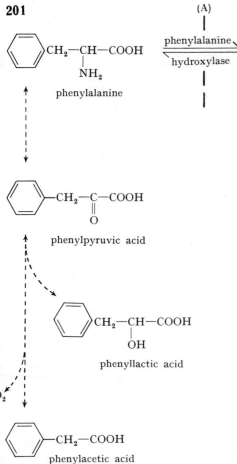
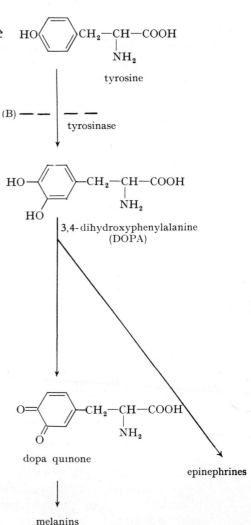

Tyrosine can also be catabolized to homogentisic acid **(202)**

202

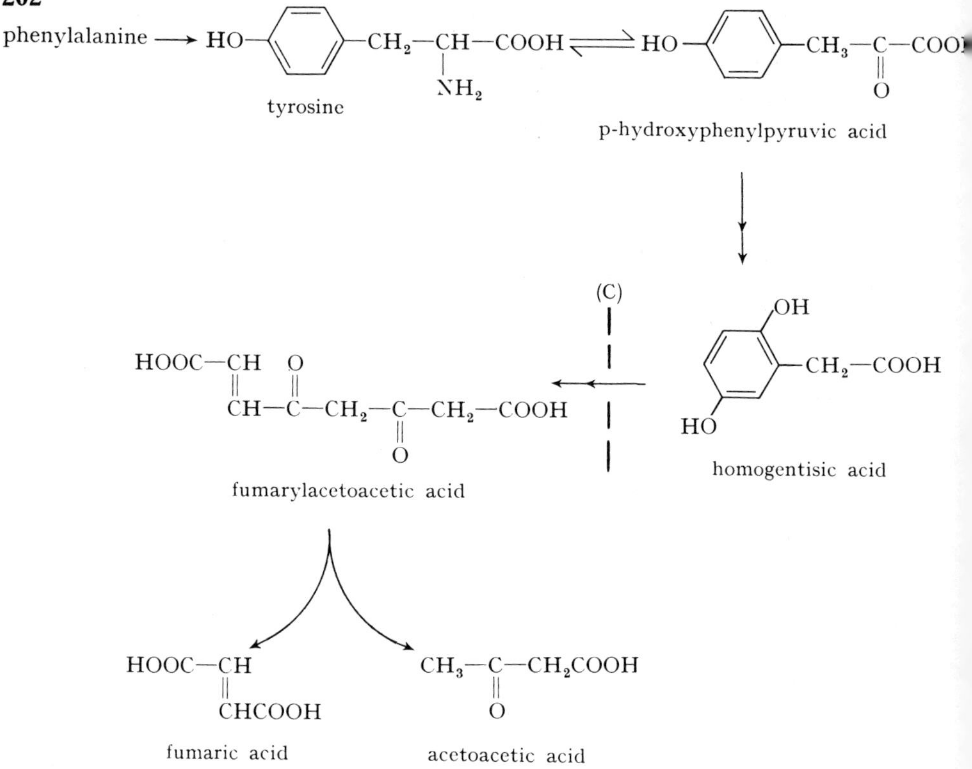

phenylalanine ⟶ HO—⟨⟩—CH$_2$—CH(NH$_2$)—COOH ⇌ HO—⟨⟩—CH$_3$—C(=O)—COOH

tyrosine   p-hydroxyphenylpyruvic acid

(C)

HOOC—CH=CH—C(=O)—CH$_2$—C(=O)—CH$_2$—COOH ← homogentisic acid (HO—⟨⟩(OH)—CH$_2$—COOH)

fumarylacetoacetic acid

↙   ↘

HOOC—CH=CHCOOH   CH$_3$—C(=O)—CH$_2$COOH

fumaric acid   acetoacetic acid

There are three well-characterized defects in the metabolism of these two amino acids. Phenylalanine that is not required for protein synthesis is normally hydroxylated to form tyrosine. In individuals who lack phenylalanine hydroxylase activity (A), the normal conversion of phenylalanine to tyrosine is prevented. Instead, phenylalanine, and phenylpyruvic, phenyllactic, and phenylacetic acids accumulate in elevated concentrations in blood and urine. This condition is called phenylketonuria. These substances are toxic to the developing central nervous system, and without proper and prompt treatment irreversible nerve system damage will result. For this reason, it is now a standard procedure to analyze the serum and urine or both of all newborns to detect phenylketonuria. Effective treatment is the maintenance of phenylalanine at nontoxic levels. There is no endogenous source for this essential amino acid, but most proteins contain phenylalanine. Therefore, to sustain a reduced intake and not diminish the supply of other amino acids, the phenylalanine is selectively removed from protein hydrolysates prior to their administration. This approach requires constant monitoring of serum phenylalanine levels to assure that there is an adequate supply for growth and development but that the level is not great enough to be toxic. If the enzyme catalyzing the oxidation of homogentisic acid is absent (C), a very rare but apparently innocuous condition called alkaptonuria develops. Homogentisic acid is excreted with the urine, which becomes dark upon standing or following the addition of alkali. The dark material is formed when homogentisic acid is oxidized to the quinone and then polymerizes.

The conversion of tyrosine to the epinephrines produces two important hormones, epinephrine and norepinephrine. This pathway is seen in Figure 9-2. They are produced in certain parts of the nervous system and by the adrenal medulla. Methylation of norepinephrine yields the 3-methoxy derivative, normetanephrine. Similarly, the methylation of epinephrine produces the 3-methoxy compound, metanephrine. Both of these compounds are further metabolized to yield 3-methoxy-4-hydroxymandelic acid. These four compounds, epinephrine, norepinephrine, metanephrine, and normetanephrine, are collectively called the catecholamines.

The rationality for calling these compounds catecholamines is apparently derived from the fact that the substitution on the benzene ring is similar to that of catechol.

203

catechol

vanillin

Further corruption of nomenclature is encountered in the case of 3-methoxy-4-hydroxylmandelic acid. This compound is frequently referred to as VMA which can be translated as vanilylmandelic acid. The excuse for using this name apparently is because the ring structure in 3-methoxy-4-hydroxylmandelic acid is somewhat similar to that of vanillin **(203)**.

The analysis of the catecholamines is usually done to detect pheochromocytomas and neuroblastomas. These are tumors of the sympathetic nervous system which occur in the adrenal medulla or other sites where there are chromaffin cells. These tumors secrete excessive amounts of epinephrine and norepinephrine, which in turn give rise to elevated quantities of the various metabolites.

The very interesting ether-amino acid, thyroxine, is also produced from tyrosine through a number of synthetic steps (Fig. 9-2).

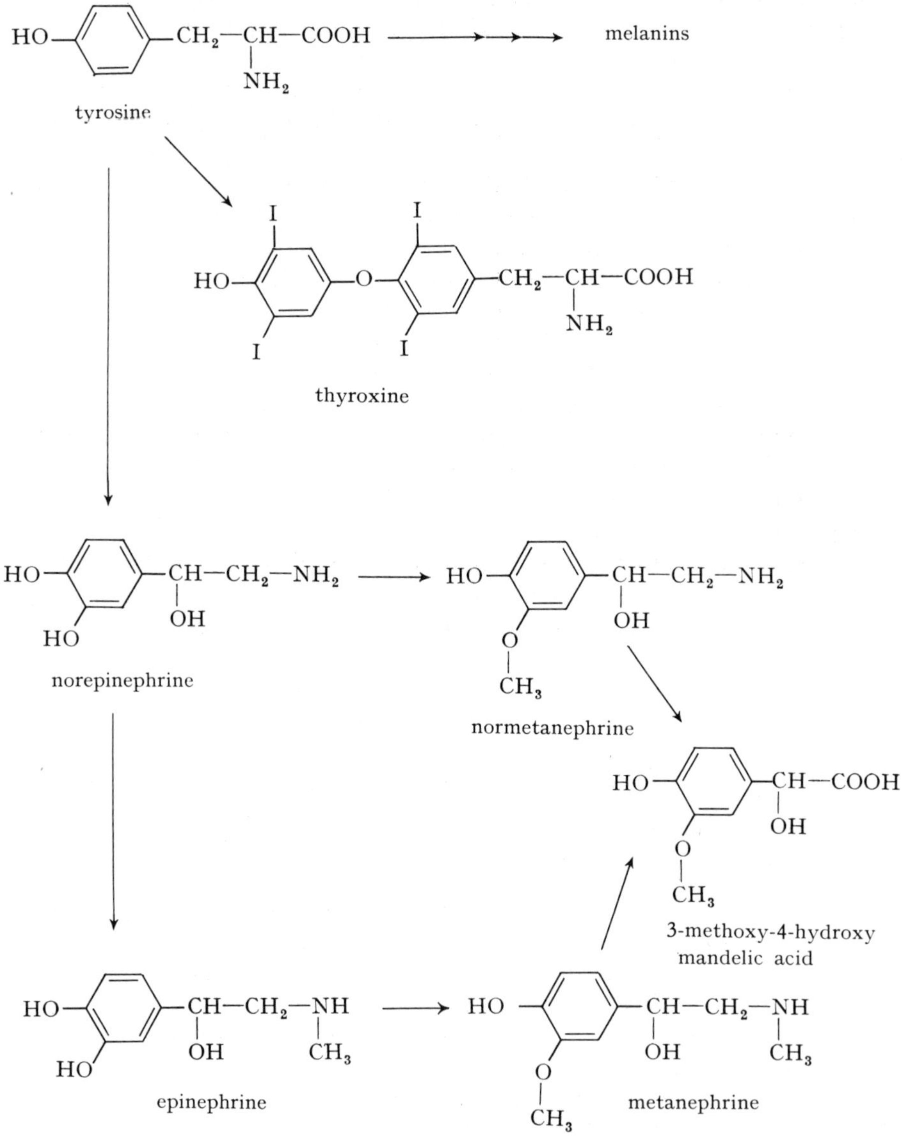

Fig. 9-2. Tyrosine metabolism.

This iodinated amino acid is a hormone produced in the thyroid glands; to a large extent it regulates the metabolic rate in mammals.

TRYPTOPHAN METABOLISM

Tryptophan can be metabolized to serotonin. The exact role and importance of serotonin is not yet resolved, but it is known that it is probably involved in blood clotting and brain function. Tumors of the argentoffin cells, called carcinoids, excrete excessive quantities of serotonin. These tumors can be detected by the increased excretion of 5-hydroxyindoleacetic acid in the urine. In laboratory parlance, this compound is commonly known as 5-HIAA.

204

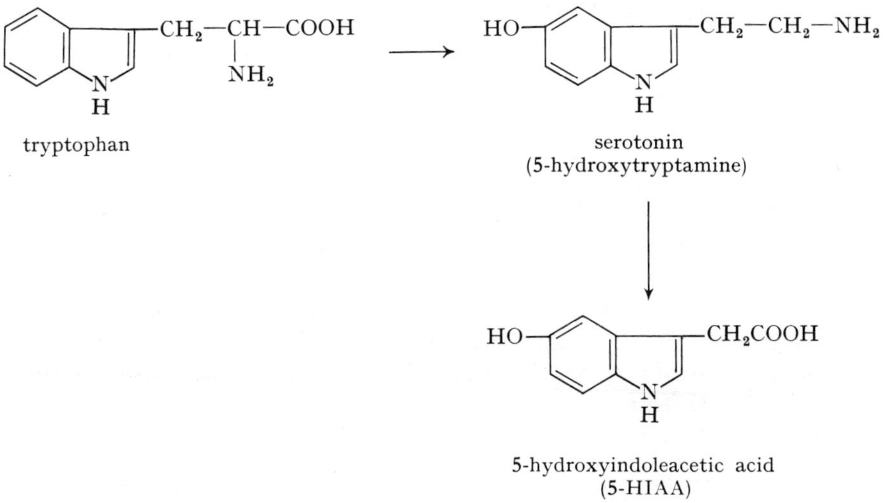

$$\text{tryptophan} \longrightarrow \text{serotonin (5-hydroxytryptamine)} \longrightarrow \text{5-hydroxyindoleacetic acid (5-HIAA)}$$

There are several other abnormalities of amino acid metabolism due to congenital enzyme deficiencies in addition to those presented here. Most of these conditions have profound effects including mental retardation and death. One of these is maple syrup disease, so named because the urine of the victims has the aroma of maple syrup. The defect is in the metabolism of the branched-chain amino acids, leucine, isoleucine, and valine (**205**).

205

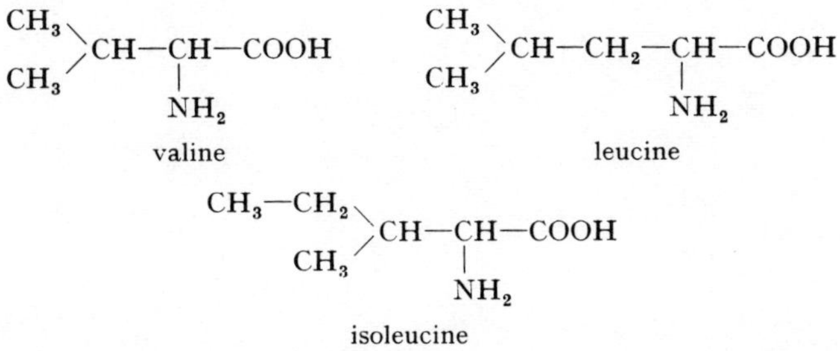

valine, leucine, isoleucine

SUMMARY OF AMINO ACID METABOLISM

To generalize about the catabolism of amino acids, they can be degraded in one of three ways. The glucogenic amino acids are those that can be converted to pyruvic acid, which in turn can be converted to glucose. The amino acids that can be converted to acetic acid and the ketone bodies are termed ketogenic. The third category of amino acids are those which yield both pyruvate and acetate. An example is phenylalanine and tyrosine, which yield acetoacetic acid and fumaric acid (202), which can be converted to acetic and pyruvic acids, respectively.

206

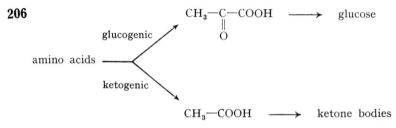

The essential amino acids cannot be synthesized or are not synthesized in adequate amounts and must be supplied in the diet. However, nonessential amino acids are synthesized by the organism in sufficient quantities to meet the needs. They can be synthesized in part by reversal of reaction (**195**) and transamination (**196**).

CHAPTER 10

NUCLEIC ACID METABOLISM

Cells contain all the information required for the reproduction of similar cells by cellular division. Germ cells, and indeed probably all cells of an organism, have sufficient information to reproduce the entire organism.

The genetic information required to assemble and organize new cells is contained in large molecules of DNA. Cells contain thousands of different proteins and the specifications ("blueprints") for their biosynthesis are contained in DNA molecules as a rather simple code. Every plant and animal has these DNA molecules grouped into structures called chromosomes. The code for each specific protein is carried in a particular region of the DNA molecule. These regions have been referred to as "genes."

One might ask why information for only protein synthesis is present and why blueprints for nonprotein structures should not also be necessary. The answer is that for practical purposes all reactions are enzymatic, and thus by dictating the kind of enzyme the kind of product is also determined.

CELLULAR DIVISION

In somatic cell division (mitosis) each daughter cell receives identical DNA molecules (genetic information). A hypothetical cell containing a single chromosome (DNA) must first duplicate the chromatin material (Fig. 10-1) in order that the daughter cell be genetically identical (diploid).

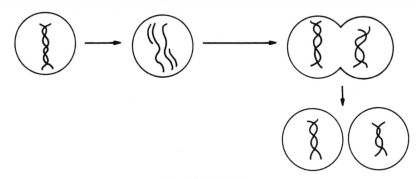

Fig. 10-1. Mitosis.

Meiosis is a special form of cell division. Here each daughter cell, ovum or spermatozoon, receives one-half the number of chromosomes (haploid). In a hypothetical case, where the ovum contains one chromosome (one strand of DNA) and the spermatozoon likewise contains one strand of DNA, the fertilized egg (Fig. 10-2) has a complete complement of chromosomes (diploid).

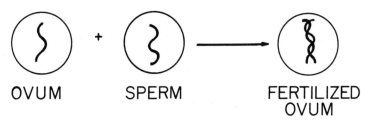

OVUM   SPERM   FERTILIZED
                                 OVUM

Fig. 10-2. Haploid cells produced by meiosis combining by fertilization to form a diploid cell.

ROLE OF NUCLEIC ACIDS

DNA's importance lies in the fact that it contains the genetic information necessary for protein synthesis. It serves as a protein pattern indirectly, however, in that an intermediate RNA molecule serves as the template for protein synthesis. As was indicated earlier, protein molecules can hydrogen bond to form a stable helical configuration. DNA forms similar structures because in nucleic acids, adenine and thymine, as well as cytosine and guanine, can hydrogen bond to form what are called base pairs.

207

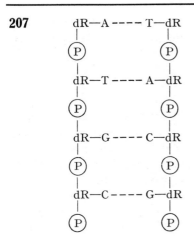

A = adenine
T = thymine
C = cytosine
G = guanine
(P) = phosphate
dR = deoxyribose

TRANSCRIPTION

When the cell needs to synthesize proteins, a strand of DNA serves as a pattern for the synthesis of a strand of RNA, called messenger RNA or mRNA. The base sequence in mRNA is dictated by the base sequence in the DNA.

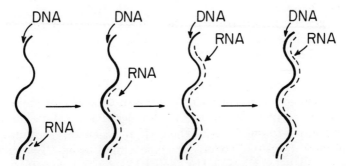

Fig. 10-3. Synthesis of messenger RNA on DNA template.

If the base sequence starting from one end of the DNA molecule is AGTC, then the base sequence in the newly synthesized mRNA molecule from the same end will be UCAG (**208**).

208

```
                Messenger
        DNA       RNA
        ┌A ─── U┐
        ├G ─── C┤
        ├T ─── A┤
        └C ─── G┘
```

The genetic information in the base sequence of the DNA is copied in a complementary fashion during the synthesis of mRNA. This process is called transcription.

GENETIC CODE

When the mRNA is complete it moves to the cytoplasm and there functions as a template for the assembly of protein molecules. There are about 20 naturally occurring amino acids, and there are only four different bases which occur in appreciable concentrations in RNA. Therefore, there are not enough different bases to code 1:1 for the 20 different amino acids. The messenger RNA cannot dictate the amino acid sequence in a protein on the basis of one base for one amino acid. If we consider a base pair or a duplex code to dictate the amino acid sequence, then the number of base combinations is 4^2 or 16. Again, the code would be insufficient to code for all the naturally occurring amino acids. If we consider a triplet code, that is one where a sequence of three bases dictate an amino acid, then there are 4^3 or 64 different code words, adequate to code for all the naturally occuring amino acids. Experimentation has shown that a triplet code is indeed used by biologic organisms, and the code is apparently universal for the mammalian and bacterial systems that have been examined.

The genetic information in the mRNA, in code form, describes the amino acid sequences of the proteins to be synthesized.

The bases on the messenger RNA are read, therefore, three at a time in a nonoverlapping fashion, and each group of three bases corresponds to a particular amino acid. Such a group is called a codon. There are 64 code words or codons, but only 20 different amino acids, so that there is an excess of codons for designating the amino acids on a simple one-to-one basis. The code is degenerate, meaning that there is more than one codon for some amino acids. For example, leucine is coded by UUG, UUA, and CUG. The degenerate codons can be likened to synonyms in our language. Other codons are probably concerned with polypeptide chain initiators and terminators. In our language this might be considered analogous to starting a sentence with a capital and ending with a period. Other codons are called nonsense, meaning it is not known what, if anything, they code for. It is reasonable to assume that many of them do have a function, perhaps analogous to various forms of punctuation in our written language.

Some examples of the base composition and sequence of codon are shown in Table 10-1.

TABLE 10-1. EXAMPLES OF BASE CODING

Triplet	Amino acid
AAA	Lysine
CCC	Proline
UUU	Phenylalanine
GGU	Glycine
GAA	Glutamic acid
UCU	Serine

AMINO ACID ACTIVATION

Some preliminary reactions must take place before messenger RNA can act as a template for protein assembly. One is the activation of amino acids through reaction with ATP. The activated amino acids then react with small soluble RNA molecules called transfer RNA (tRNA). Each amino acid has at least one specific transfer RNA with which it combines. The amino acid tRNA complexes combine with mRNA as directed by base pairing. When transfer units are joined to messenger units in the proper sequence, protein synthesis takes place.

PROTEIN SYNTHESIS

The site of protein synthesis in the cytoplasm is on ribosomes. Ribosomes are cytoplasmic structures that contain roughly equal amounts of RNA and protein and exist in several states of aggregation. It used to be believed that the ribosomal RNA was the template for protein synthesis, but now it seems that the ribosomes operate as a carrier for the messenger RNA.

For protein synthesis, several individual ribosomes become attached to a more or less fully extended, singly stranded, mRNA molecule forming arrays of ribosomes which have been called polyribosomes or polysomes. During actual protein synthesis, a single ribosome attaches to one of the terminal ends of the mRNA and travels along it while peptide bonds are being formed on the complex. As the ribosome moves down the mRNA molecule, a growing polypeptide chain results until the completed protein is synthesized. The protein is synthesized one amino acid at a time starting from the N terminus.

The addition of each sequential amino acid to the chain requires, along with the necessary enzymes and cofactors, the tri complex association of ribosome-mRNA-tRNA-amino acid. Each tRNA carries an anticodon that can form a temporary bond with one of the codons in the messenger RNA, thus bringing the proper amino acid into proper sequence in the polypeptide.

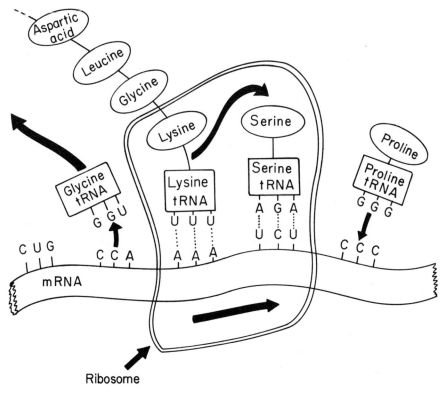

Fig. 10-4. Protein synthesis.

The process of protein synthesis is described in Figure 10-4. At the left glycine tRNA is seen leaving the mRNA after having brought a molecule of glycine for incorporation. Just to the right of that is seen a lysine tRNA still attached to the mRNA and the peptide chain which contains the terminal lysine it transported to the site. To the immediate right of that is a serine tRNA with serine attached. The next step in the protein synthesis is the transfer of the chain to the serine moiety of the serine-serine-tRNA-complex. At the far right is a proline-proline-tRNA-molecule becoming attached to the mRNA. Proline will follow serine as the next amino acid in the chain.

TRANSLATION

There are many questions about protein synthesis still to be answered. How are the correct three bases chosen out of a group of many? Consider the following six bases on a strand of mRNA.

209

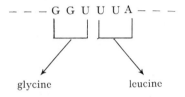

They can be decoded as glycine and leucine; but if read differently

210

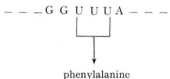

a different amino acid is involved. Protein synthesis is apparently very free of errors so that the mechanism for correct reading must be a good one. Another interesting aspect is the movement of the ribosome along the mRNA. This is locomotion and is usually associated with muscular activity. The mechanism for this motion is not understood.

MUTATION

Mutations result in differences in protein structure due to changes in the genetic code. Radiation (ultraviolet, x-ray, γ-ray, etc.) is known to induce mutations through destruction or alteration of the bases, which changes the DNA code. Chemicals which react with the bases can also cause mutations.

Sickle cell anemia is a condition characterized by the presence of hemoglobin S. Hemoglobin S differs from hemoglobin A, the normal adult hemoglobin, by only a single amino acid. Hemoglobin S has valine in the place of glutamic acid of hemoglobin A. The code for glutamic acid is GAA and the triplet for valine is GUU. In this instance, as in most, the difference in the codon is only a single base — A to a U — but the effects are profound.

NUCLEIC ACID SYNTHESIS

The synthesis of the nucleic acids requires the presence of an RNA or DNA molecule as a pattern or "primer." As indicated above in cellular division, the chromatin DNA serves as a pattern for the synthesis of new chromatin DNA. The base sequence of the new DNA is determined by base pairing, and the ends of the old and new strands are reversed (antiparallel).

211

```
            old           new
           — A           T —
           — C           G —
           — G           C —
           — T           A —

           — A           T —
           — C           G —
           — G           C —
           — T           A —
```

RNA can be synthesized with a DNA template, as described above for mRNA synthesis, or from other RNA molecules. DNA synthesis requires the deoxynucleoside triphosphates and RNA synthesis demands the nucleoside triphosphates (**212**).

212

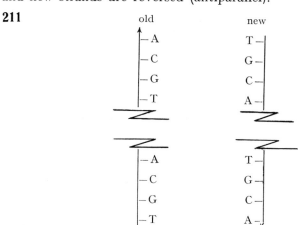

PURINE AND PYRIMIDINE METABOLISM

There are no dietary requirements for purines and pyrimidines since they are readily synthesized by the mammalian organism. However, those consumed in the diet are in part used in the synthesis of nucleic acids.

Pyrimidine catabolism is complete; they are metabolized to CO_2, water, and ammonia. The purines, however, are not metabolized completely; they leave a residue of uric acid.

213

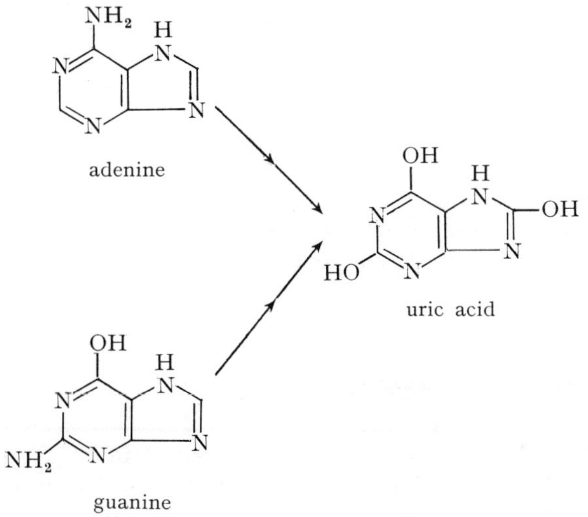

adenine

uric acid

guanine

Compounds containing adenine and guanine are degraded to release the bases, which are converted to uric acid.

Normally, the uric acid level in serum is between 2 and 6 mg/dl. Abnormal purine catabolism occurs in a condition called gout. The defect in this inflammatory disease is lack of proper regulation of purine synthesis so that excessive quantities are produced. Uric acid is not very soluble, and levels are easily reached at which it begins to form crystalline deposits. One common site of crystalline precipitation is in the joint of the big toe.

CHAPTER 11

COMMON METABOLIC PATHWAYS

Thus far, discussions of metabolism have been concerned with reactions largely characteristic of particular types of compounds. Here, reactions common to the products of those metabolic schemes will be considered.

ACETYL COENZYME A

Acetyl coenzyme A is often referred to as the key intermediate of metabolism. The reason is that catabolism of carbohydrates, fats, and amino acids results in the formation of acetyl coenzyme A.

The catabolism of carbohydrates as presented leads to the production of lactic acid. Further metabolism of lactic acid results in the formation of acetyl coenzyme A.

214
$$CH_3-\underset{\underset{\text{lactic acid}}{|}}{\overset{OH}{C}H}-COOH \quad \xrightarrow[]{NAD^+ \quad\quad NAD\cdot H + H^+} \quad CH_3-\underset{\underset{\text{pyruvic acid}}{}}{\overset{O}{\overset{\|}{C}}}-COOH$$

215
$$CH_3-\underset{\underset{O}{\overset{\|}{C}}}{C}-COOH \xrightarrow[CO_2 \quad\quad CoASH]{NAD^+ \quad\quad NAD\cdot H + H^+} CH_3-\underset{\underset{O}{\overset{\|}{C}}}{C}-S-CoA$$

pyruvic acid acetyl coenzyme A

The oxidation of pyruvic acid to acetyl CoA is rather complex and is shown in an abbreviated form.

The amino acids are catabolized by one of two routes.

216

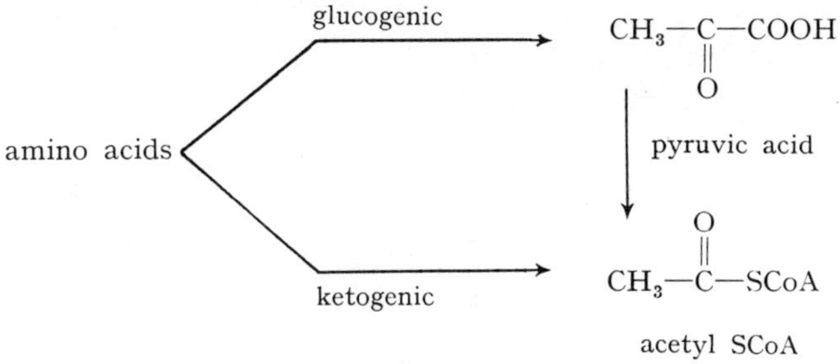

$$\text{amino acids} \begin{cases} \xrightarrow{\text{glucogenic}} CH_3-\underset{\underset{O}{\|}}{C}-COOH \\ \\ \xrightarrow{\text{ketogenic}} \end{cases}$$

pyruvic acid

$$\downarrow$$

$$CH_3-\underset{\underset{O}{\|}}{C}-SCoA$$

acetyl SCoA

The pyruvic acid produced from the degradation of the glucogenic amino acids is converted to acetyl CoA by reaction **215**.

Fatty acids are also converted to acetyl CoA.

217   fatty acids $\longrightarrow$ $CH_3-COSCoA$
                                    acetyl CoA

The key role played by acetyl coenzyme A can be summarized as

218

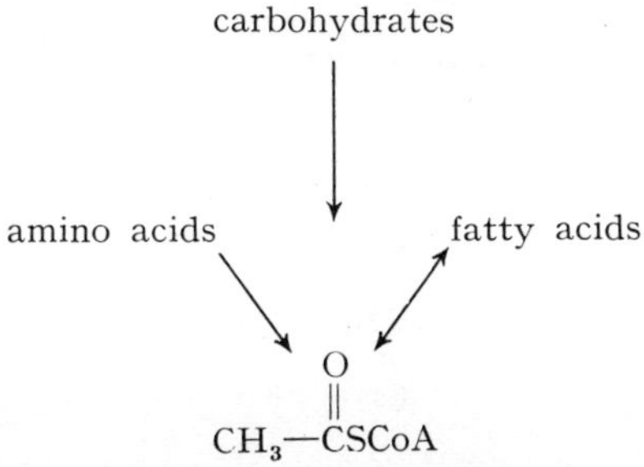

carbohydrates

amino acids $\searrow$ $\swarrow$ fatty acids

$$CH_3-\underset{\underset{O}{\|}}{C}SCoA$$

Notice that the conversion of carbohydrates and amino acids to acetyl CoA is an irreversible process. The only reversible process involving acetyl CoA is the synthesis of long-chain fatty acids.

TRICARBOXYCYLIC ACID CYCLE

When large amounts of energy are required, the tissues are able to convert carbohydrates, amino acids, and fatty acids to acetyl CoA and further oxidize them in what is known as the tricarboxycylic acid cycle (Fig. 11-1). The net overall reaction taking place in this

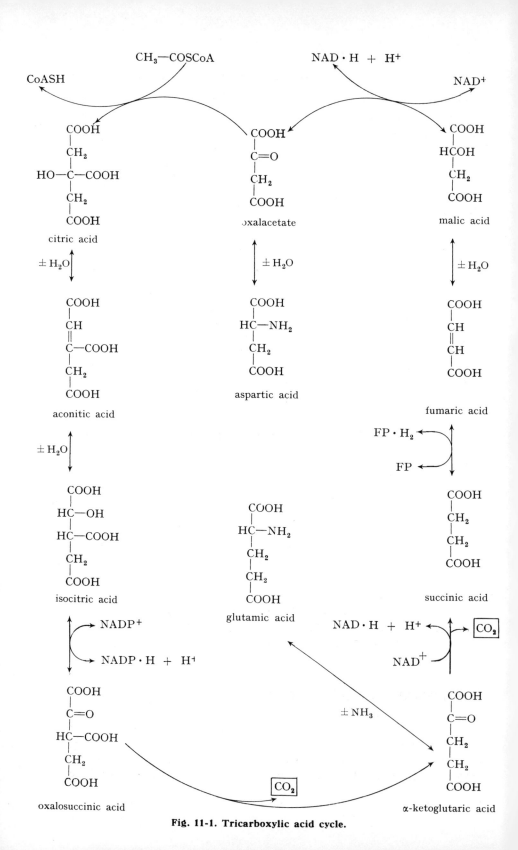

Fig. 11-1. Tricarboxylic acid cycle.

cycle is the complete oxidation of acetic acid to CO_2, protons, and electrons. Notice that acetic acid is oxidized with just one passage through this extremely efficient cyclic process. The compound required for the initial condensation with acetyl CoA is oxalacetic acid; aspartic acid can be deaminated to yield oxalacetic acid. However, glutamic acid can be used to start the cycle when oxidatively deaminated or converted in a transamination reaction to yield α-ketoglutaric acid.

Fats cannot be metabolized well unless there is concomitant carbohydrate metabolism, thus the origin of the phrase "fats burn in carbohydrate's flame." This is because the oxalacetic acid required for the initial condensation with acetyl coenzyme A is supplied largely from pyruvic acid.

219 $\quad CH_3-\underset{\underset{O}{\|}}{C}-COOH + CO_2 \longrightarrow HOOC-CH_2-\underset{\underset{O}{\|}}{C}-COOH$

$\qquad$ pyruvic acid $\qquad\qquad\qquad\qquad$ oxalacetic acid

If, however, pyruvic acid cannot be produced from carbohydrates, as in diabetes mellitus or in starvation where there is no glycogen available, then the only source of oxalacetic acid is from the degradation of amino acids, such as aspartic and glutamic acid or both (Fig. 11-1). This explains in part why diabetes mellitus is a wasting disease. Individuals in this situation cannot utilize carbohydrate as a source of oxalacetic acid and must rely on their amino acids, which in turn are derived from proteins.

ELECTRON TRANSPORT SYSTEM

Oxidation of metabolites in the tricarboxycylic acid cycle and in the other metabolic reactions presented earlier result in reduced coenzymes. These coenzymes are reoxidized by participation in the electron transport systems. At the start of the electron transport system shown in Figure 11-2, oxidation of metabolites (substrates, SH_2) result in oxidized metabolites (S) and reduced coenzymes ($NAD \cdot H + H^+$). The $NAD \cdot H + H^+$ then reacts with flavoprotein (FP) and is oxidized while the flavoprotein is reduced (FP $\cdot H_2$). $NADP \cdot H$ can be oxidized by a similar process. The flavoprotein reacts with a cytochrome, (cyto $\cdot Fe^{+++}$) to become reoxidized as the cytochrome is reduced (cyto $\cdot Fe^{++}$). At this step the protons are excluded. The electrons are passed along to several different cytochromes, by a series of similar oxidation-reduction reactions, until

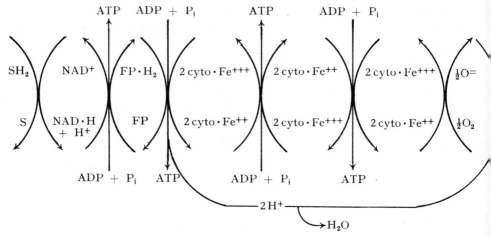

Fig. 11-2. Electron transport system.

the final cytochrome. The last cytochrome in the series reduces molecular oxygen, and the oxygen ion reacts with the proton released earlier to produce water.

$$2H^+ + O^= \longrightarrow H_2O$$

The electron transport system is important for several reasons: (1) it explains how oxygen finally enters into metabolism; (2) it provides a mechanism for the regeneration (reoxidation) of oxidation-reduction coenzymes; (3) it provides the majority of the energy derived from metabolic processes. Simultaneously with the flow of electrons through the system there is formation of ATP from P_i and ADP. The processes whereby ATP is formed from ADP + P_i, called oxidative phosphorylation, are poorly understood. The series of coupled reactions in the electron transport system, starting with the reduced coenzymes to the reduction of oxygen, represent decreasing energy levels. The chemical energies released in the various reactions are utilized to form the high-energy ATP from ADP + P_i.

CHAPTER 12

CLINICALLY IMPORTANT ENZYMES

The assay of enzyme activities in biologic specimens has become one of the most important functions of clinical chemistry laboratories. The particular enzyme assays performed by a given laboratory will depend upon several factors, including the types of patients served and the requirements of the attending staff.

Presented here are very brief descriptions of eight enzymes which are most commonly assayed in clinical laboratories.

AMYLASE

The enzyme amylase hydrolyzes the linear portions of starch, amylose, to maltose.

221
$$\text{amylose} \xrightarrow{\text{amylase}} \text{maltose}$$

Normally, serum contains small amounts of amylase activity. The origin of the enzyme in normal serum has not been established. It is known that the salivary glands and pancreas, which produce amylase for digestive purposes, release it into circulation in certain situations.

Amylase activity can be assayed in one of three general ways. One method, the saccharogenic technique, measures the amount of maltose released per unit of time, usually measured as reducing substances, expressed as glucose equivalents. Another assay method, the amyloclastic technique, measures the disappearance of the starch substrate. Usually this is done by measuring the disappearance of the starch-iodine color with time.

The third method for measuring amylase activity is to employ a starch substrate that is covalently linked to a dye. As the starch is hydrolyzed, smaller molecules containing the dye are released. At the end of the incubation period an organic solvent is added to the reaction mixture. The unhydrolyzed starch is precipitated, but the smaller molecules remain in solution The amount of dye in the supernatant solution can be measured colorimetrically, and it is directly related to enzyme activity.

Normal values, expressed as units of activity, are dependent upon the assay method. Therefore, no exact values for normal individuals will be given, but most techniques have normal values of less than 200 units.

An increased serum amylase activity occurs in mumps, which affects the salivary glands. Serum and urine amylase activity assays are most useful in suspected cases of pancreatitis. In acute pancreatitis amylase activity values can be five times or more the normal values.

LIPASE

Lipase, an enzyme of pancreatic origin, is also released into circulation in abnormally high concentrations in inflammatory conditions of the pancreas. The assay of lipase activity can be accomplished by titration of the fatty acids produced when the serum is incubated with a suitable triglyceride substrate.

222

$$\begin{array}{c} CH_2O-\overset{O}{\overset{\|}{C}}-R_1 \\ | \\ CHO-\overset{O}{\overset{\|}{C}}-R_2 \\ | \\ CH_2O-\overset{O}{\overset{\|}{C}}-R_3 \end{array} + 3\,H_2O \xrightarrow{\text{lipase}} \begin{array}{c} CH_2OH \\ | \\ CHOH \\ | \\ CH_2OH \end{array} \begin{array}{c} R_1COOH \\ + \\ R_2COOH \\ + \\ R_3COOH \end{array}$$

A more sensitive technique for measuring lipase activity is to measure the fatty acids released colorimetrically as copper soaps. The serum is incubated with a buffered triglyceride emulsion, and the reaction is stopped by the addition of chloroform. The chloroform extracts the fatty acids. An aqueous solution of a cupric salt is equilibrated with the chloroform solution. The fatty acids form chloroform-soluble copper soaps. Next, a colorless chelating agent (diethyldithiocarbamate) is added which chelates the copper and becomes colored. A photometric measurement of the color is then a measure of the fatty acids, which in turn is directly related to enzyme activity. The same reagents and procedure can be used to measure serum fatty acid concentrations except that no substrate or incubation is employed. Human pancreatic lipase apparently is quite specific for the long-chain triglycerides and will not hydrolyze small, simple triglycerides such as tributyrin or triacetin. A common substrate for lipase activity is olive oil.

The lipase activity is expressed in units which are related to microequivalents or micromoles of fatty acids released from the substrate under the conditions of the assay.

Following the onset of pancreatitis, increased serum amylase and lipase activities appear at different rates (Fig. 12-1).

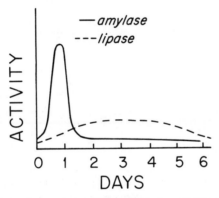

Fig. 12-1. Time-course of serum amylase and lipase activities as a result of acute pancreatitis.

There is a prompt increase in amylase activity which disappears from the serum in a matter of 18 to 24 hours. Lipase activity increases much more slowly and remains elevated for a longer period of time. Because of this difference, the assay of lipase is often extremely important in detecting those cases of pancreatitis which the physician sees sometime after the onset.

SERUM GLUTAMATE OXALACETATE TRANSAMINASE

During the past few years the enzyme, serum glutamate oxalacetate transaminase (SGOT) has received considerable attention and assumed great importance clinically. SGOT is quantitated in the clinical laboratory by measuring the oxalacetate formed during the following reaction.

223

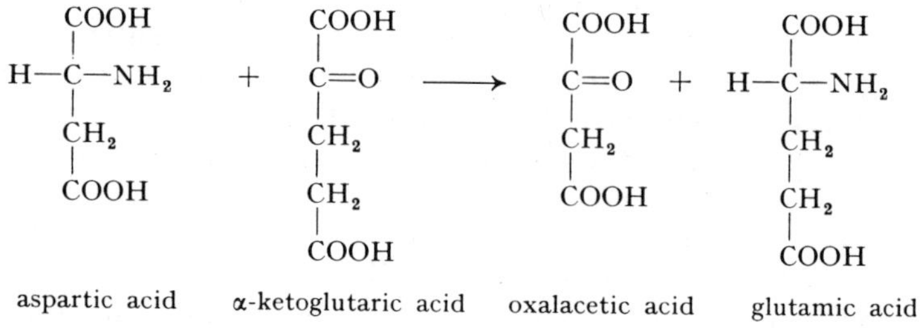

$$\underset{\text{aspartic acid}}{\begin{array}{c}\text{COOH}\\|\\\text{H}-\text{C}-\text{NH}_2\\|\\\text{CH}_2\\|\\\text{COOH}\end{array}} + \underset{\alpha\text{-ketoglutaric acid}}{\begin{array}{c}\text{COOH}\\|\\\text{C}=\text{O}\\|\\\text{CH}_2\\|\\\text{CH}_2\\|\\\text{COOH}\end{array}} \longrightarrow \underset{\text{oxalacetic acid}}{\begin{array}{c}\text{COOH}\\|\\\text{C}=\text{O}\\|\\\text{CH}_2\\|\\\text{COOH}\end{array}} + \underset{\text{glutamic acid}}{\begin{array}{c}\text{COOH}\\|\\\text{H}-\text{C}-\text{NH}_2\\|\\\text{CH}_2\\|\\\text{CH}_2\\|\\\text{COOH}\end{array}}$$

The oxalacetate reacts with a stable aromatic diazonium salt to form a formazan-like derivative which is measured colorimetrically. The assay can be standardized by the use of oxalacetate as the primary standard.

224

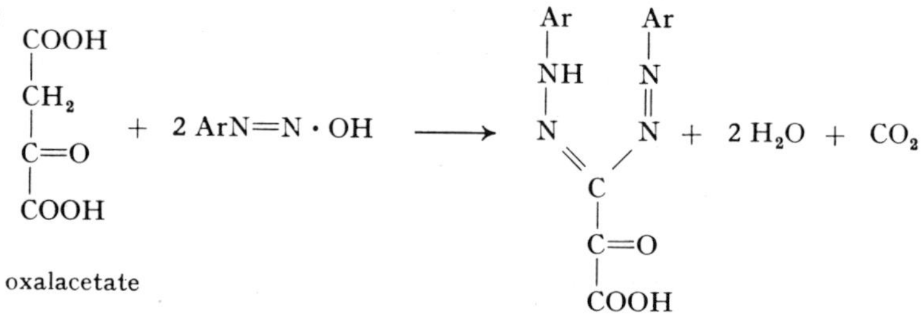

$$\underset{\text{oxalacetate}}{\begin{array}{c}\text{COOH}\\|\\\text{CH}_2\\|\\\text{C}=\text{O}\\|\\\text{COOH}\end{array}} + 2\,\text{ArN}=\text{N}\cdot\text{OH} \longrightarrow \begin{array}{c}\text{Ar}\quad\text{Ar}\\|\quad\;\;|\\\text{NH}\quad\text{N}\\|\quad\;\;\|\\\text{N}\quad\;\;\text{N}\\\diagdown\;\diagup\\\text{C}\\|\\\text{C}=\text{O}\\|\\\text{COOH}\end{array} + 2\,\text{H}_2\text{O} + \text{CO}_2$$

In the assay described, the unit of enzyme activity is equivalent to the amount of enzyme in 1 liter of serum that will form 25 μmoles of oxalacetate per minute. Normal values range between 5 and 25 units.

The measurement of SGOT is important in liver and heart diseases since both organs are rich in the enzyme. In cases of myocardial infarction or liver damage the enzyme is released in abnormal quantities into the circulation. For example, in extensive liver necrosis, the SGOT activity may exceed 2,000 units.

CREATINE PHOSPHOKINASE

Another enzyme important in laboratory medicine is creatine phosphokinase (CPK, creatine kinase). CPK equilibrates creatine phosphate and ADP to form ATP and creatine.

225

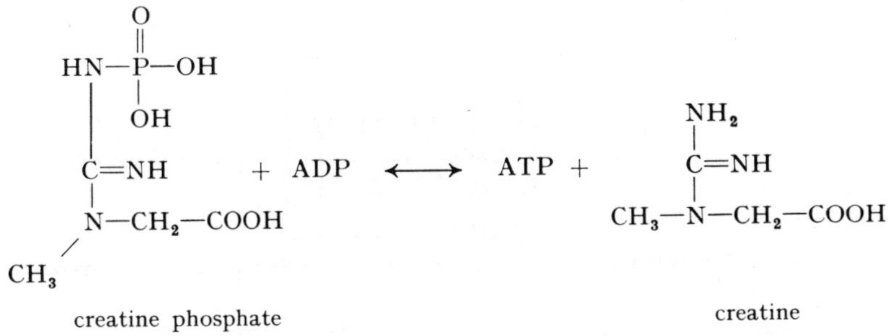

$$\underset{\text{creatine phosphate}}{\begin{array}{c} \text{HN}-\overset{\overset{\text{O}}{\|}}{\underset{|}{\text{P}}}-\text{OH} \\ \text{OH} \\ \text{C}=\text{NH} \\ | \\ \text{N}-\text{CH}_2-\text{COOH} \\ / \\ \text{CH}_3 \end{array}} + \text{ADP} \longleftrightarrow \text{ATP} + \underset{\text{creatine}}{\begin{array}{c} \text{NH}_2 \\ | \\ \text{C}=\text{NH} \\ | \\ \text{CH}_3-\text{N}-\text{CH}_2-\text{COOH} \end{array}}$$

The reaction is reversible, but in most laboratories the reaction is allowed to proceed from left to right as seen here.

The creatine produced during the incubation period is measured colorimetrically by reaction with diacetyl and α-naphtol. Each unit of CPK activity is equivalent to the formation of 1 μmole of creatine from creatine phosphate per milliliter of serum per hour at 37° C.

Creatine phosphokinase activity is limited exclusively to muscle, therefore it only becomes elevated in heart or other muscle disorders. For this reason the measurement of CPK activity is important in differentiating heart and liver disease. Since SGOT is elevated in both liver and heart disease the use of both SGOT and CPK assays distinguishes heart and liver problems.

LACTATE DEHYDROGENASE

The enzyme lactate dehydrogenase (LD) is widely distributed throughout body tissues, and elevations in total serum enzymatic activity is related to a number of diseases.

226

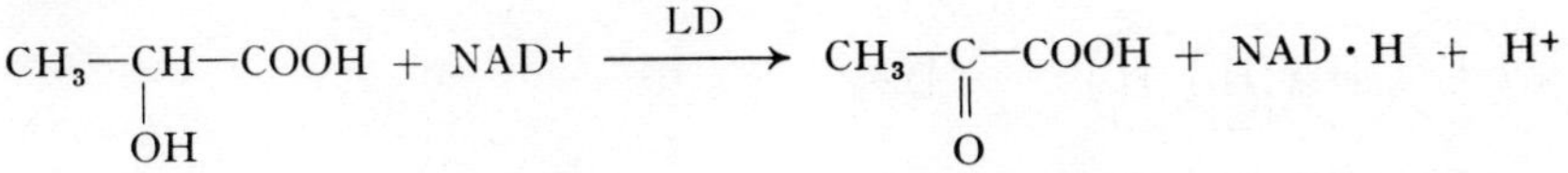

$$\text{CH}_3-\underset{\underset{\text{OH}}{|}}{\text{CH}}-\text{COOH} + \text{NAD}^+ \xrightarrow{\text{LD}} \text{CH}_3-\underset{\underset{\text{O}}{\|}}{\text{C}}-\text{COOH} + \text{NAD} \cdot \text{H} + \text{H}^+$$

The reaction catalyzed by the enzyme can be quantitated by measurement of the appearance of reduced NAD (NAD·H) by ultraviolet absorption at 340 nm. Conversely, the disappearance of lactate or appearance of pyruvate can also be measured colorimetrically.

As previously stated, elevation in total LD activity is not specific for any particular disease. More definitive information can be obtained by electrophoretic separation of the LD isozymes (Fig. 12-2).

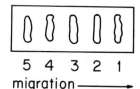

5 4 3 2 1
migration ⟶

Fig. 12-2. Electrophoretic separation of lactate dehydrogenase isozymes.

When serum containing LD is separated electrophoretically, five different isozyme fractions are obtained. The isozymes are reacted with various chemicals to yield colored areas on the electrophoretic strip which can be quantitated. Fraction 1 has its origin primarily in the heart, whereas fraction 5 comes mostly from the liver. Increased activity of fraction 1 suggests cardiac muscle necrosis, whereas increased fraction 5 activity suggests malfunction of the liver.

LD was the first enzyme shown to consist of several different isozymes. The term isozyme refers to enzymes with similar molecular weights, activities, and structures which have two or more common subunits (Fig. 12-3).

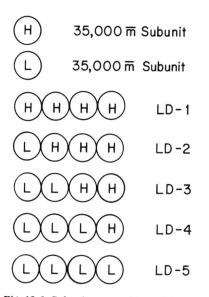

Fig. 12-3. Subunit composition of the five serum lactate dehydrogenase isozymes.

LD has two polypeptide subunits, one called H and the other L. Fraction 1 consists of four H subunits, while fraction 5 contains four L subunits. Fractions 2,3, and 4 have various combinations of H and L subunits.

CLINICAL RESPONSE OF SGOT, CPK, AND LD

The relative response of SGOT, CPK, and LD in myocardial infarction and in acute liver disorders such as hepatitis or obstruction present a typical pattern that offers useful diagnostic information. Consider first the time response of the three enzymes in myocardial infarction (Fig. 12-4). In a typical situation, CPK activity increases

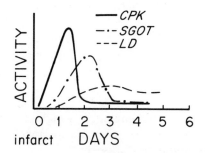

Fig. 12-4. Time-course of three serum enzyme activities as a result of myocardial infarct.

first, followed by SGOT, and sometime later by LD. Electrophoretic separation of LD isozymes would demonstrate increased activity in fraction 1.

A typical pattern for the enzyme response in an acute liver disorder would appear as shown in Figure 12-5.

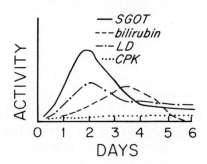

Fig. 12-5. Changes in activity of three serum enzymes and changes in bilirubin concentrations following an acute liver disorder.

Usually within a day or two the transaminase activity reaches a maximum; this peak is followed closely by lactate dehydrogenase. CPK remains essentially normal throughout. Fractionation of LD would show a predominance of activity in fraction 5. Enzymatic activity usually stays elevated for longer periods of time in liver disorders than in myocardial infarction.

The response of bilirubin is considerably slower than that of the enzymes. This means that the enzymes are much more sensitive indicators of liver disorders than serum bilirubin levels.

Frequently following a myocardial infarct the patient will develop congestive failure. This can often be detected by fractionation of the LD isozymes. As a result of the infarction fraction 1 will be elevated, but when congestive failure intervenes, fraction 5 will also become elevated. Apparently this is due to the fact that blood circulation in congestive failure is inadequate, causing liver hypoxia, which leads to the elevation in fraction 5.

PHOSPHATASES

The phosphatases are enzymes which hydrolyze phosphate esters. In clinical laboratories the phosphatase enzymes ordinarily measured are phosphomonoesterases.

227
$$R-O-\underset{\underset{OH}{|}}{\overset{\overset{O}{\|}}{P}}-OH + H_2O \longrightarrow R-OH + HO-\underset{\underset{OH}{|}}{\overset{\overset{O}{\|}}{P}}-OH$$

A number of different substrates have been used to measure phosphatase activity. The first commonly used substrate was β-glycerol phosphate.

228
$$\underset{\text{β-glycerol phosphate}}{\overset{CH_2OH}{\underset{CH_2OH}{|}}{H-C-O-(P)}} \longrightarrow \underset{\text{glycerol}}{\overset{CH_2OH}{\underset{CH_2OH}{|}}{CHOH}} + P_i$$

In the reaction, the substrate is hydrolyzed to yield glycerol and inorganic phosphate. Quantitation is accomplished by the measurement of phosphate released by the enzymolysis. The technique is complicated by the fact that serum initially contains inorganic phosphate. Therefore, two analyses must be made—one to determine the initial phosphate level, another to determine the initial phosphate plus that released from the substrate. The difference between the two analytic values provides a measure of phosphate released from the substrate. This is a cumbersome procedure, and in cases where the amount of serum inorganic phosphate present initially is high and the amount released by the enzyme is low accurate measurements are difficult.

In choosing a substrate, it is desirable to have a reaction product which does not occur naturally in biologic systems. This avoids the necessity of two assays. In a reaction using phenylphosphate as a substrate, both phenol and inorganic phosphorus are released.

229

$$\text{phenylphosphate} \longrightarrow \text{phenol} + P_i$$

Phenol can be measured colorimetrically, and since phenol does not occur in appreciable amounts in biologic systems it is not necessary to measure phenolic compounds initially present in serum.

Attempts have also been made to use substrates which yield products that are chromophores.

230

$$p\text{-nitrophenylphosphate} \xrightarrow{} p\text{-nitrophenol} + P_i$$

$$\xrightarrow{NaOH} iso\text{nitrophenol}$$

*Para*nitrophenylphosphate when hydrolyzed by phosphatase, yields *para*nitrophenol. When the solution is alkalinized, the colored product, *iso*nitrophenol, can be readily measured colorimetrically. Another chromogenic substrate, phenolphthalein monophosphate, yields phenolphthalein when hydrolyzed.

231

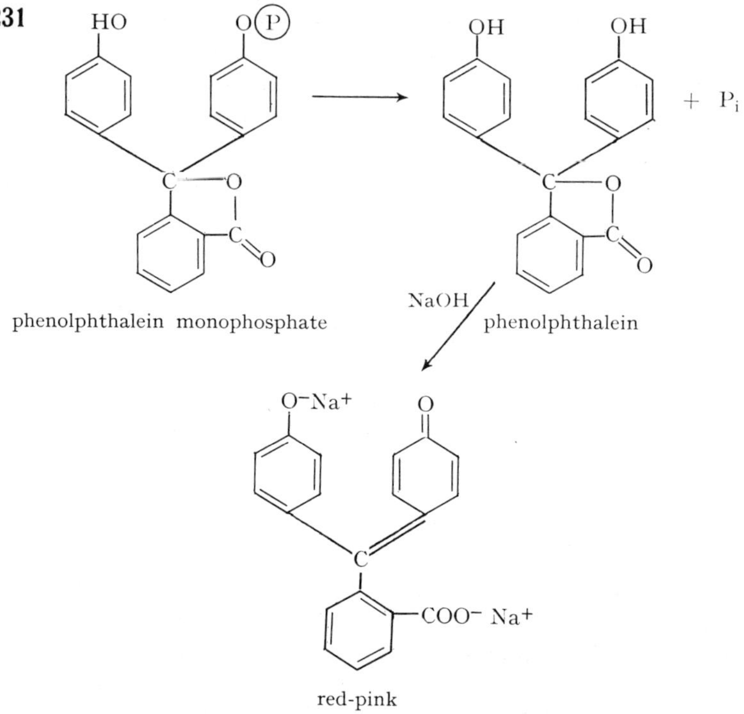

phenolphthalein monophosphate → phenolphthalein + P_i

NaOH ↓

red-pink

Upon alkalinization the familiar red-pink phenolphthalein color can be measured by ordinary colorimetric techniques. The four substrates presented here are only representatives of the various types that have been and continue to be employed in the clinical measurement of phosphatase activity.

ALKALINE AND ACID PHOSPHATASES

If serum phosphatase activities are measured at different pH values and plotted, activity versus pH, two different activities are observed (Fig. 12-6).

These different activities are not due to the same enzyme but rather represent two different types of enzymes. Those with maximum activity around pH 4 to 5 are acid phosphatases, whereas those with maximum activity around pH 10 are alkaline phosphatases. Notice that between the two peaks there is no activity, so that when acid phosphatase activity is measured, alkaline phosphatase activity does not interfere and vice versa. In practice the same substrates can be used to measure both; the only difference is the buffer and the pH used during the incubation.

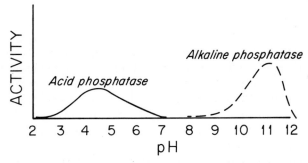

Fig. 12-6. pH-activity relationships of acid and alkaline phosphatases.

Some precautions must be observed in measuring acid phosphatase activity. For one thing, erythrocytes contain considerable acid phosphatase activity, and thus sera obtained from hemolyzed samples are not suitable for assay. Also, acid phosphatase is not stable under usual laboratory conditions, but the stability can be increased by maintaining the sample in an ice bath until it is analyzed. However, the length of time that the enzyme activity can be preserved at ice bath temperature is limited.

Acid phosphatase is widely distributed in tissues. Acid activity is found in bone, liver, kidney, and red cells. Normal serum contains a small amount of acid phosphatase activity, and most of this enzyme apparently originates in the liver and spleen. The primary value in measuring acid phosphatase activity is in the diagnosis of metastasizing prostatic carcinoma. The prostate gland contains very large amounts of acid phosphatase, and in prostatic carcinoma acid phosphatase is often released into the serum and gives rise to elevated phosphatase activities. The activities may be increased ten or more times the normal level. This is not an entirely reliable indicator, since an appreciable percentage of metastasizing prostatic carcinomas do not show elevated acid phosphatase activities. Certain other diseases, particularly those involving bone disorders, may give rise to slight elevations in acid phosphatase activity.

Measurement of alkaline phosphatase activity is usually done to detect malfunctions of liver and bone. Normal serum contains small amounts of alkaline phosphatase activity, and the origin of this enzyme is apparently bone. Normally, children show higher alkaline phosphatase activity than do adults. Increases in alkaline phosphatase activity are associated with diseases of the skeleton, such as rickets, Paget's disease, and so on. Also, liver disorders give rise to increased serum activity. Alkaline phosphatase activity is quite stable in serum, and the serum can be stored at least overnight under refrigerator conditions without appreciable loss of activity.

ACTIVITY UNITS

The normal activity units for the phosphatases are dependent upon the method used to determine them. This depends not only on the particular substrate but also the buffer and the concentrations of both. Because there is a variety of techniques to measure phosphatase activity, attempts have been made to correlate the activity as measured by one technique to that measured by another. In any given circumstance or in any given sample a conversion factor can be found. However, if one attempts to do this with a number of different samples there does not seem to be a correlation. The reason for this is that the activity observed for any given biologic sample is a composite of activity of several different phosphatase enzymes. These apparently are not isozymes but simply different enzymes with phosphatase activity.

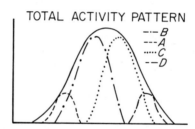

Fig. 12-7. Representation of alkaline phosphatase activity showing that it is the total of the activities of four different alkaline phosphatases.

In the example shown in Figure 12-7 the total activity is due to the additive activities of four enzymes (A, B, C, and D). It has been demonstrated that the substrate response, pH optimum, and so on for the various phosphatases differ. Thus, in moving from one biologic sample to another, where the relative proportions of the enzymes vary, it would not be surprising that the correlation between two different substrates and two different buffers or both would be quite different. Thus, to be professionally honest, there is no way that one can convert phosphatase units using one substrate to that of another.

ALDOLASE

Serum aldolase is an enzyme which is occasionally assayed in the laboratory.

232

fructose-1,6-di-P →(aldolase) dihydroxyacetone phosphate + 3-phosphoglyceraldehyde

$$\text{fructose-1,6-di}P \xrightarrow{\text{aldolase}} \begin{array}{c} CH_2O\text{-}P \\ | \\ C=O \\ | \\ CH_2OH \end{array} + \begin{array}{c} CHO \\ | \\ CHOH \\ | \\ CH_2O\text{-}P \end{array}$$

Fructose-1,6-diphosphate is cleaved to yield dihydroxyacetone phosphate and 3-phosphoglyceraldehyde. After incubation, alkali is added to hydrolyze the phosphate esters to dihydroxyacetone and glycerin.

233

$$\begin{array}{c} CH_2O\text{-}P \\ | \\ C=O \\ | \\ CH_2OH \end{array} \xrightarrow{\text{NaOH}} \begin{array}{c} CH_2OH \\ | \\ C=O \\ | \\ CH_2OH \end{array} + P_i$$

dihydroxyacetone-P → dihydroxyacetone

The products of this reaction are converted to colored derivatives by reaction of 2,4-dinitrophenylhydrazine to form the corresponding hydrazones.

234

$$\begin{array}{c} CH_2OH \\ | \\ C=O \\ | \\ CH_2OH \end{array} + NH_2NH\text{-}\phi(NO_2)_2 \longrightarrow \begin{array}{c} CH_2OH \\ | \\ C=N\text{-}NH\text{-}\phi(NO_2)_2 \\ | \\ CH_2OH \end{array} + H_2O$$

dihydroxyacetone   2,4-dinitrophenylhydrazine   hydrazone

Glyceraldehyde-P reacts in a similar fashion (**235**). The glyceraldehyde also reacts with 2,4-dinitrophenylhydrazine to form a hydrazone.

235

$$\begin{array}{c} CHO \\ | \\ CHOH \\ | \\ CH_2O\text{-}P \end{array} \xrightarrow{\text{NaOH}} \begin{array}{c} CHO \\ | \\ CHOH \\ | \\ CH_2OH \end{array} + P_i$$

glyceraldehyde

The distribution of aldolase activity is very much like lactate dehydrogenase in that it occurs in most tissues. Elevations in serum aldolase activity can be due to a number of causes. For instance, it is elevated in prostatic carcinoma, myopathies, acute myocardial infarction, acute liver disease, and many other disease states. Erythrocytes also contain adolase, and thus serum obtained from hemolyzed blood samples is not suitable for assay.

CHAPTER 13

EQUILIBRIUM, ACIDS, BASES, AND BUFFERS

Understanding the independent actions and interactions of acids, bases, and buffers is essential to the comprehension of many chemical reactions.

EQUILIBRIUM

Consider the reaction where two reactants A and B interact to form the products C and D.

236 $\qquad$ A + B $\longleftrightarrow$ C + D

The rate of the reaction—that is, the rate at which C and D accumulate and the rate at which A and B disappear—is graphically illustrated in Figure 13-1.

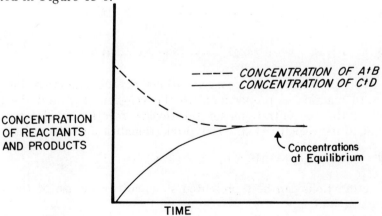

Fig. 13-1. Time-course of reaction A + B $\rightleftarrows$ C + D as it approaches equilibrium.

Starting with only A and B, the rate of the reverse reaction (where A and B are formed from C and D) will be zero initially because there is no C and D. As the concentrations of C and D increase with time, the rate of the reverse reaction increases. Given enough time, the rate of the forward reaction will equal that of the reverse reaction and at this point the reaction system is in equilibrium.

For the reaction to take place, reactants A and B must come in contact with each other, or, in other words, collide. This is analogous to the situation involving enzymes.

237 $\qquad$ A + B $\longrightarrow$ A-B $\longrightarrow$ C + D

A and B come together to form a reactive complex which decomposes to form C and D. Originally the concentrations of A and B are high. There is therefore a greater probability for collision, resulting in a high reaction rate for the forward reactions. The conversion of A and B to C and D reduces the concentration of the reactants and the probability of their collision, and the forward reaction velocity decreases. With the formation of increased amounts of C and D, the probability of their collisions resulting in conversion to A and B increases and so then does the reverse reaction rate.

REACTION VELOCITIES

The reaction velocity in both directions is proportional to the reactant concentrations. We can label the forward reaction velocity as V_1, and the reverse reaction velocity as V_2

238 $\qquad$ A + B $\xrightarrow{V_1}$ C + D

239 $\qquad$ C + D $\xrightarrow{V_2}$ A + B

The product of the concentrations of the reactants for the forward reaction is proportional to the velocity (V_1), and the product of the concentrations for the reverse reaction is also proportional to its velocity (V_2). The proportionalities can be written as

240 $\qquad$ [A] [B] $\propto V_1$ $\qquad$ [C] [D] $\propto V_2$

These functions can be represented as equalities by use of the constants k_1 and k_2.

241 $\qquad$ [A] [B] $k_1 = V_1$ $\qquad$ [C] [D] $k_2 = V_2$

The purpose of these constants is to introduce appropriate dimensions and also a numerical value such that the product of the concentrations times the factor does indeed equal the observed velocity. This numerical value is determined experimentally. The reactants A and B are expressed in molarity (moles/liter), and velocity in this example will be moles/hour/liter. The product of A and B is moles^2/liters^2, and the correct dimensions of the velocity constant (k) would be 1/mole/liter/hour.

EQUILIBRIUM CONSTANT

At equilibrium, the rate of the forward reaction must be equal to the reverse reaction.

242
$$V_1 = V_2$$

Therefore

$$[A]\ [B]\ k_1 = [C]\ [D]\ k_2$$

Now by dividing both sides of the equation by [A] [B] and then dividing by k_2, the following expression is derived

243
$$\frac{k_1}{k_2} = \frac{[C]\ [D]}{[A]\ [B]}$$

A constant divided by a constant is equal to a constant. Thus

244
$$\frac{k_1}{k_2} = K$$

K is defined as the equilibrium constant for a reaction and represents the ratio of $\frac{k_1}{k_2}$.

245
$$K = \frac{[C]\ [D]}{[A]\ [B]}$$

If K is large this means that when the reaction reaches equilibrium there will be more C and D than there will A and B, and the forward reaction predominates. If K were to equal one, it would mean that equilibrium is reached when [A] [B] is equal to [C] [D]. If K is small, that would mean that little A and B are converted to C and D. Therefore, from a value of K one is able to predict to what extent a reaction will take place.

EFFECTS OF REACTANT CONCENTRATION

Consider now a practical application of the equilibrium equation where it is possible to determine the amount of the products formed. The reaction begins with 1 mole/liter each of acetic acid and ethyl alcohol and proceeds until equilibrium is reached. The concentration of the ethyl acetate and water formed will be represented as x, so that at equilibrium we have remaining $(1 - x)$ mole/liter of acetic acid and ethyl alcohol.

At equilibrium

246
$$\overset{1-x}{\text{HOAc}} + \overset{1-x}{\text{EtOH}} \rightleftharpoons \overset{x}{\text{EtOAc}} + \overset{x}{\text{H}_2\text{O}}$$

The equilibrium constant of this particular reaction has been found to be 4. By substituting these numerical values, the concentration of the products can be determined.

247
$$K = \frac{[\text{EtOAc}][\text{H}_2\text{O}]}{[\text{HOAc}][\text{EtOH}]} = 4$$

$$4 = \frac{[x][x]}{[1-x][1-x]}$$

$$4 = \frac{x^2}{1 - 2x + x^2}$$

$$4(1 - 2x + x^2) = x^2$$

$$4 - 8x + 4x^2 = x^2$$

$$4 - 8x + 3x^2 = 0$$

$$x = 2,\ 2/3$$

$$x = 0.667\ M = [\text{EtOAc}] = [\text{H}_2\text{O}]$$

(NOTE: the number 2 is greater than the concentration initially and therefore is not an applicable solution for the equation.)

Stated another way, at equilibrium one-third of the acetic acid and ethyl alcohol will remain whereas two-thirds will be converted to the products ethyl alcohol and water.

Using the same approach and increasing the concentration of acetic acid to $2\ M$, a slightly different solution is obtained. Substituting the numerical values into the equation, rearranging, and solving as before reveals that the amount of product formed is $0.833\ M$.

248
$$\underset{[\text{HOAc}]}{2-x} + \underset{[\text{EtOH}]}{1-x} \rightleftharpoons \underset{[\text{EtOAc}]}{x} + \underset{[\text{H}_2\text{O}]}{x}$$

$$4 = \frac{[x][x]}{[2-x][1-x]}$$

$$4 = \frac{x^2}{2 - 3x + x^2}$$

$$8 - 12x + 4x^2 = x^2$$

$$8 - 12x + 3x^2 = 0$$

$$x = 0.833 \, M = [\text{EtOAc}] = [\text{H}_2\text{O}]$$

Repeating the same process this time increasing the concentration of acetic acid to 10 M provides a value of x which is equal to 0.967 M.

249
$$\underset{[\text{HOAc}]}{10-x} + \underset{[\text{EtOH}]}{1-x} \rightleftharpoons \underset{[\text{EtOAc}]}{x} + \underset{[\text{H}_2\text{O}]}{x}$$

$$4 = \frac{[x][x]}{[10-x][1-x]}$$

$$4 = \frac{x^2}{10 - 11x + x^2}$$

$$40 - 44x + 4x^2 = x^2$$

$$40 - 44x + 3x^2 = 0$$

$$x = 0.967 \, M = [\text{EtOAc}] = [\text{H}_2\text{O}]$$

Not only does the equilibrium equation predict the yield of products, it also demonstrates the law of mass action. By increasing the concentration of one reactant, the concentrations of the products are increased; similarly, by decreasing the concentration of one or more reactants, less product is formed.

To obtain the greatest possible yield, reactant concentrations can be increased. Also, the same results can be obtained by removing one of the products. If there were a device for removing water or ethyl acetate from the previous reaction, accumulation of more product would result because it would disrupt the equilibrium. This technique of forcing reactions to completion by using large concentrations of reactants is common in clinical methods, particularly colorimetric procedures. Most colorimetric procedures use an excess of one or more of the reactants in order to yield the greatest quantity of colored material. This leads to greater sensitivity for colorimetric procedures.

DISSOCIATION

The same type of reasoning is applicable when applied to dissociation

250
$$\text{HA} \underset{V_2}{\overset{V_1}{\longleftrightarrow}} \text{H}^+ + \text{A}^-$$

HA, a hypothetical acid, dissociates to produce protons and anions. The forward reaction velocity, V_1, equals k_1 (velocity constant) times the concentration of HA; and similarly, V_2 equals k_2 times the concentration of protons and anions.

251
$$V_1 = k_1 \text{[HA]}$$
$$V_2 = k_2 \text{[H}^+\text{] [A}^-\text{]}$$

At equilibrium
$$V_1 = V_2$$

And substituting yields
$$k_1 \text{[HA]} = k_2 \text{[H}^+\text{] [A}^-\text{]}$$

If the equation is first divided by [HA] then by k_2, the formula can be written as

252
$$\frac{k_1}{k_2} = \frac{\text{[H}^+\text{] [A}^-\text{]}}{\text{[HA]}} = K$$

K (k_1 divided by k_2) represents the equilibrium constant and is usually called the dissociation constant denoting the type of reaction involved.

Knowing the dissociation constant, the hydrogen ion and anion concentrations can be calculated. Accordingly, if

253
$$K = 10^{-6}$$
$$\text{[HA]} = 10^{-2} \, M$$
$$\text{[H}^+\text{]} = \text{[A}^-\text{]} = x \text{ at equilibrium}$$

Therefore, the concentration of undissociated HA at equilibrium will be equal to the total amount added to the solution minus that which dissociates to produce protons and anions: $\text{[HA]} = 10^{-2} \, M - x$ at equilibrium. The hydrogen ion and anion concentrations are going to be small with respect to the initial concentration of HA because the dissociation constant is small and little HA will dissociate. Thus, the assumption can be made that the concentration of HA at equilibrium, $[HA]_e$ is equal to the initial concentration $10^{-2} M$.

254
$$[HA]_e = [HA]_{initial} - [H^+] \approx [HA]_{initial} = 10^{-2} M$$

Writing the dissociation equation, substituting the values, and solving, the hydrogen ion concentration yields a value of 10^{-4} molar.

255
$$K = \frac{[H^+][A^-]}{[HA]}$$

$$10^{-6} = \frac{[H^+]^2}{10^{-2}}$$

$$[H^+]^2 = 10^{-8}$$

$$[H^+] = \sqrt{10^{-8}} = 10^{-4} M$$

The solution to the equation indicates that our assumption that the hydrogen ion concentration is small with respect to the initial acid concentration is correct.

256
$$[HA]_e = [HA]_{initial} - [H^+] = 10^{-2} - 10^{-4} =$$
$$0.01 - 0.0001 = 0.0099 \approx 0.01 M$$

HYDROGEN ION CONCENTRATION AND pH

Hydrogen ion concentration is a commonly used quantity in the sciences, particularly chemistry. It is often very difficult to speak of or deal with hydrogen ion concentrations, since the numbers are very small and the problem of locating the decimal is difficult to say the least. A technique for expressing hydrogen ion concentration which eliminates some of these difficulties is the convention of pH. By definition

257
$$pH \equiv \log \frac{1}{[H^+]} \text{ or } pH = -(\log [H^+])$$

A note of caution should be introduced here in that there is no such thing as a negative log. The negative sign merely indicates that the expression is a reciprocal. From the previous example, the hydrogen ion concentration is $10^{-4} M$, or, expressed another way, is pH 4.

258
$$pH = \log \frac{1}{10^{-4}} = \log 1 - \log 10^{-4} = 0 - (-4) = 4$$

Neutrality is defined as the situation where the hydrogen ion concentration equals the hydroxyl ion concentration. An acid solution is one where the hydrogen ion concentration is greater than the hydroxyl ion concentration; likewise, an alkaline or basic solution is one where the hydroxyl ion concentration exceeds the proton concentration.

259
$[H^+] = [OH^-]$ = neutral
$[H^+] > [OH^-]$ = acidic
$[OH^-] > [H^+]$ = basic or alkaline

ACIDS AND BASES

An acid will be defined here as a proton donor.

260
$$\underset{\text{acid}}{HA} \rightleftharpoons H^+ + \underset{\text{base}}{A^-}$$

$$BOH \rightleftharpoons B^+ + \underset{\text{base}}{OH^-}$$

$$OH^- + H^+ \rightleftharpoons H_2O$$

HA, an acid, is indeed a proton donor because it dissociates to provide protons and anions. If a base is defined as a proton acceptor, then A^- is a base — i.e., in the reverse reaction it combines with a proton to yield HA. A base is usually thought of as yielding a hydroxyl group. If it is defined as a proton acceptor, then how does a base such as NaOH function? It functions because the hydroxyl ion produced by the dissociation of NaOH combines avidly with protons to produce water.

DISSOCIATION OF WATER

Water is a very interesting substance — it functions both as an acid and as a base.

261
$$\underset{\text{acid}}{H_2O} \rightleftharpoons H^+ + \underset{\text{base}}{OH^-}$$

$$\underset{\text{base}}{H_2O} + H^+ \rightleftharpoons \underset{\text{acid}}{H_3O^+}$$

Such substances which act as both acid and base are called amphoteric substances or ampholytes. An expression of dissociation can be written for water in the same way as any other substance.

262 $$H_2O \longleftrightarrow H^+ + OH^-$$

$$K = \frac{[H^+][OH^-]}{[H_2O]}$$

$$[H_2O] = 1{,}000 \text{ g}/18 \text{ m̄} \approx 55\ M$$

The concentration of pure water is about 55 molar. The density of water is essentially one, and one liter would contain 1,000 g. The molecular weight is 18, yielding a concentration of about 55 molar. In the dissociation of water very little will be converted to protons and hydroxyl ions. Therefore, at equilibrium the concentration of water is essentially equal to the initial or total concentration of water. For practical purposes, this value of 55 molar is constant.

263 $$[H_2O]_{\text{initial}} = [H_2O]_{\text{equilibrium}} - [H^+] \approx [H_2O]_{\text{initial}} \simeq 55\ M$$

Since K is a constant, 55 is also essentially a constant, the product of the two will be designated as K_w.

264 $$K = \frac{[H^+][OH^-]}{[H_2O]}$$

$$K \cdot [H_2O] = [H^+][OH^-]$$

$$K_w = [H^+][OH^-] = 10^{-14}$$

K_w equals the hydrogen ion concentration times the hydroxyl ion concentration; by experiment this is found to be equal to about 10^{-14}. This value is affected by temperature, but for practical use 10^{-14} is a valid number.

The hydrogen ion concentration equals the hydroxyl ion concentration of water, since there is only one source for both ions; so then pure water should be a neutral solvent.

265 $$[H^+] = [OH^-] = \text{neutral solution}$$

In practice, however, the pH of water is apt to be about 6. The reason for this is that atmospheric CO_2 can react with water to produce carbonic acid, which can dissociate yielding protons and bicarbonate ions making the solutions slighlty acidic.

266 $$CO_2 + H_2O \rightleftharpoons H_2CO_3 \rightleftharpoons H^+ + HCO_3^-$$

pOH is defined the same way that pH is defined.

267
$$\mathrm{pOH} = \log \frac{1}{[OH^-]} \text{ or } \mathrm{pOH} = -(\log [OH^-])$$

The dissociation of pure water provides a neutral solution with $[H^+] = [OH^-]$

268
$$K_w = 10^{-14} = [H^+][OH^-]$$
$$[H^+] = [OH^-] \text{ neutral solution}$$
$$10^{-14} = [H^+]^2$$
$$[H^+] = 10^{-7}$$
$$\mathrm{pH} = 7$$
$$\mathrm{pOH} = 7$$

Solving the equation reveals that the hydrogen ion concentration is 10^{-7} and the pH is 7. Since the hydrogen ion concentration equals the hydroxyl ion concentration, the latter concentration will also be 10^{-7} or pOH 7. The sum of pH + pOH is 14. It can be shown that the pH plus the pOH is always equal to 14. This is a convenient equation to remember for checking answers when solving pH problems.

COMMON ION EFFECT

One thing which affects the dissociation of a substance is the presence of a common ion. This can be best illustrated by an example. Assume that there is an acid, HA, at $10^{-2}\,M$. To this solution a similar concentration of the sodium salt of this acid, NaA, is added. Further, let the dissociation constant for HA equal 10^{-6}. HA will dissociate to yield protons and anions as in **254**, but this time an extra source of A^- is supplied from NaA. Sodium salts are for the most part extremely soluble and essentially completely dissociated. Writing expressions for the dissociation of HA in the presence of NaA appears as before

269
$$K = \frac{[H^+][A^-]}{[HA]}.$$

The concentration of HA at equilibrium will be equal to the total amount of HA added, minus that which dissociates. Since the dissociation is very small (dissociation constant of 10^{-6}), for practical purposes $[HA]_e$ is equal to the concentration of HA initially added.

270 (a) $[HA]_{equilibrium} = [HA]_{initial} - [H^+] \approx [HA]_{initial} \approx 10^{-2} M$

(b) $[A^-]_{equilibrium} = [NaA]_{initial} + [A^-]_{HA} \approx [NaA]_{initial} \approx 10^{-2} M$

The concentration of A^- at equilibrium will be equal to the NaA concentration plus the A^- formed from the dissociation of HA. Similarly, $[A^-]$ formed from the dissociation of HA is very small so that the concentration of A^- is essentially the same as the initial concentration of NaA. Substituting these values into **269** gives a hydrogen ion concentration of $10^{-6} M$ or pH 6.

271
$$10^{-6} = \frac{[H^+]\,[10^{-2}]}{[10^{-2}]}$$

$$[H^+] = 10^{-6} M$$

$$pH = 6$$

Compare this answer with that obtained in **255** where the HA concentration and the dissociation constant were the same and the pH was 4. In the presence of NaA, the pH has been increased to 6. Saying it another way, the hydrogen ion concentration has been reduced 100-fold. This illustrates the effect of a common ion, where the concentration of A^- is supplied almost exclusively from the NaA

272
$$HA \rightleftharpoons H^+ + \underset{\underset{NaA}{\uparrow}}{A^-}$$

The presence of a large amount of A^- represses the ionization of HA; thus the amount of hydrogen ion produced by the dissociation of HA is decreased, as evidenced by the increase in pH value in the above example.

BUFFERS

The common ion factor has a great application in buffer systems. A pH buffer system consists of a weak acid and a salt of a weak acid; it resists changes in pH. The titration curves where pH is plotted against milliliters of base added for solutions of HCl and acetic acid appear as shown in Figure 13-2. With small increments of NaOH added, the HCl has a sharp change in pH at the equivalence point, from approximately pH 3 to pH 10. The titration curve of acetic acid, a weak acid, is quite different. As opposed to HCl, the pH of acetic acid is initially greater, and the change throughout the curve is less per unit of NaOH added. The area of the curve where there

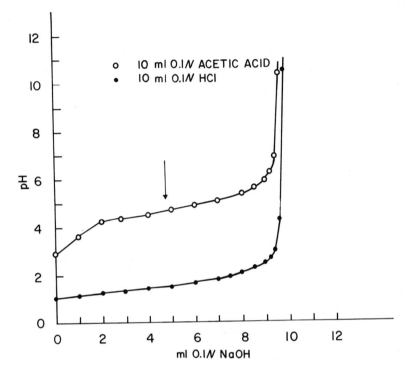

Fig. 13-2. Titration curves for acetic and hydrochloric acids. *Arrow* indicates the midpoint of the titration of acetic acid and the area of smallest change in pH.

is the least change in pH per milliliters of NaOH added is midway between the start of the titration and the end point, as indicated by the arrow; this is the area of greatest buffer effect. With an appreciable addition of NaOH solution, the pH in this region remains fairly constant. This represents the action of a buffer solution, because at this point the solution consists of acetic acid and sodium acetate, thus fulfilling the requirements of a buffer—a solution of a weak acid and a salt of the weak acid. The mechanism of buffering action can be seen by the following example.

273   HOAc + NaOH $\longrightarrow$ NaOAc + H_2O
        NaOAc + HCl $\longrightarrow$ NaCl + HOAc

With an acetate buffer, the acetic acid will react with bases, such as NaOH, to produce sodium acetate and water. The buffer reacts with the strong base to form a very weak one. Similarly, if a strong acid, such as HCl, is added to the acetate buffer, the sodium acetate reacts to produce NaCl, a neutral salt, and acetic acid, a weak acid. Thus a weak acid results from the reaction of the strong acid with AcO$^-$

DISSOCIATION CONSTANTS AND BUFFERS

The hydrogen ion concentration of a buffer system can be calculated from the dissociation expression of the weak acid concerned. Consider for example, a buffer system containing the hypothetical weak acid, HA, and its sodium salt.

274. (a)
$$HA \rightleftharpoons H^+ + A^-$$
$$NaA \longrightarrow Na^+ + A^-$$

At equilibrium, the dissociation expression for the weak acid may be written as

(b) $$K = \frac{[H^+][A^-]}{[HA]}$$

For practical purposes, the sodium salt NaA is completely dissociated, resulting in an infinitely large K value; therefore, a similar expression for NaA is meaningless. Both sides of the dissociation equation are divided by $[H^+]$ and K (**274** (c) and (d), respectively).

(c) $$\frac{K}{[H^+]} = \frac{[A^-]}{[HA]}$$

(d) $$\frac{1}{[H^+]} = \frac{1}{K} \times \frac{[A^-]}{[HA]}$$

The equation can be written in log form.

(e) $$\log \frac{1}{[H^+]} = \log \frac{1}{K} + \log \frac{[A^-]}{[HA]}$$

By definition

$$pH \equiv \log \frac{1}{[H^+]}$$

$$pK \equiv \log \frac{1}{K}$$

Therefore, both pH and pK can be substituted into equation **274** (e) for their respective equalities

(f) $$pH = pK + \log \frac{[A^-]}{[HA]}$$

In this form, the expression is known as the Henderson-Hasselbalch equation.

K and pK or both values can be readily found in most chemistry handbooks. If they are not available they can be readily determined experimentally. Consider a buffer solution where the concentration of [HA] is equal to that of [A$^-$], then pH equals pK. To determine pK experimentally, a solution is prepared so that [A$^-$] = [HA]. The pH of the solution is determined with a pH meter.

275 $\qquad$ [HA] = [A$^-$]

$$\text{pH} = \text{p}K + \log \frac{[\text{A}^-]}{[\text{HA}]} = \text{p}K + \log 1 = \text{p}K + 0$$

By using the Henderson-Hasselbalch equation, it can be demonstrated that dilution has little or no effect on a buffer pH. Substituting, HA with pK 6 and a concentration of 0.01 M and NaA with a concentration of 0.10 M, the calculated pH would be 7. If the concentration of NaA and HA were increased 10-fold, the resulting pH would also be 7 (**276** (c)). Conversely, by decreasing the concentration, the pH would be the same (**276** (d)).

276

(a) $\qquad \text{pH} = 6 + \log \dfrac{[\text{A}^-]}{[\text{HA}]}$

(b) $\qquad \text{pH} = 6 + \log \dfrac{0.1}{0.01} = 6 + \log 10 = 6 + 1 = 7$

(c) $\qquad \text{pH} = 6 + \log \dfrac{1.0}{0.1} = 6 + \log 10 = 6 + 1 = 7$

(d) $\qquad \text{pH} = 6 + \log \dfrac{0.001}{0.0001} = 6 + \log 10 = 6 + 1 = 7$

It can be concluded that a buffer system does not change pH with dilution. In practice, this is true providing the dilution is not greater than 10-fold.

Using the Henderson-Hasselbalch equation, the amounts of salt and acid required to prepare an inorganic phosphate buffer, pH 7.1, can be calculated. In this case, the weak acid is supplied by sodium dihydrogen phosphate (monobasic sodium phosphate) yielding $H_2PO_4^-$ which dissociates to provide the protons. The pK for this reaction is 6.8. The salt is provided by disodium hydrogen phosphate (dibasic sodium phosphate) and dissociates completely to yield the $HPO_4^=$ radical (**277** (c)). Substituting and solving leads to a value of 2 for the ratio of $HPO_4^=$ to $H_2PO_2^-$ (**277** (d)).

277 (a) $\quad NaH_2PO_4 \longrightarrow Na^+ + H_2PO_4^-$

(b) $\quad H_2PO_4^- \rightleftharpoons H^+ + HPO_4^= \quad pK = 6.8$

(c) $\quad Na_2HPO_4 \longrightarrow 2\,Na^+ + HPO_4^=$

(d) $\quad pH = pK + \log \dfrac{[HPO_4^=]}{[H_2PO_4^-]}$

$\quad\quad 7.1 = 6.8 + \log \dfrac{[HPO_4^=]}{[H_2PO_4^-]}$

$\quad\quad 0.3 = \log \dfrac{[HPO_4^=]}{[H_2PO_4^-]}$

$\quad\quad \text{antilog } 0.3 = 2 = \dfrac{[HPO_4^=]}{[H_2PO_4]}$

This means that for a phosphate buffer, pH 7.1, the concentration of $HPO_4^=$ needs to be twice that of the $H_2PO_4^-$.

BUFFER CONCENTRATIONS

In practice, buffers are usually described in terms of concentration. For instance, it is possible to calculate the number of grams of NaH_2PO_4 and Na_2HPO_4 required to make a 0.1 M inorganic phosphate buffer, pH 7.1. When buffer concentration is stated it indicates the concentration of both the salt and acid forms. Thus, the concentration of $[HPO_4^=] + [H_2PO_4^-]$ is 0.1 M. As seen in **(277 (d))**, the two occur in a ratio of 2 parts salt to one part acid **(278 (a))**. Substituting the equivalent of $HPO_4^=$, 2 $[H_2PO_4^-]$, the concentration of $H_2PO_4^-$ can be calculated **(278 (b))**. Solving for the concentration of $HPO_4^=$ yields a value of 0.0666 M **(278 (c))**.

278 (a) $\quad\quad [HPO_4^=] = 2\,[H_2PO_4^-]$

(b) $\quad\quad 2\,[H_2PO_4^-] + [H_2PO_4^-] = 0.1\,M$

$\quad\quad\quad\quad 3\,[H_2PO_4^-] = 0.1\,M$

$\quad\quad\quad\quad [H_2PO_4^-] = 0.0333\,M$

(c) $\quad\quad [HPO_4^=] + 0.0333\,M = 0.1\,M$

$\quad\quad\quad\quad [HPO_4^=] = 0.0666\,M$

Knowing the molarity and the specific compounds involved, it is simple to calculate the number of grams needed to prepare the desired volume of buffer solution.

CHAPTER 14

ACID-BASE BALANCE

All biologic systems have very definite pH limits in which they can survive. In the human, if the blood pH remains below 7.35 or above 7.45 for very long, disastrous results ensue.

SOURCES OF ACIDS

Metabolic systems are acid producers. One very large source is CO_2, which becomes hydrated to form carbonic acid, which dissociates to release protons.

279
$$CO_2 + H_2O \rightleftharpoons H_2CO_3$$
$$H_2CO_3 \rightleftharpoons H^+ + HCO_3^-$$

When sulfur-containing amino acids are metabolized, the sulfur released is the equivalent of H_2SO_4.

280
$$\text{S-aa's} \longrightarrow H_2SO_4$$

Also when organic phosphorus compounds such as nucleic acids are degraded, they release the equivalent of phosphoric acid.

281
$$\text{P(organic)} \longrightarrow H_3PO_4$$

Carbohydrate metabolism leads to glyceric, pyruvic, and lactic acids — good examples of organic metabolic acids. Of course, in diseases affecting carbohydrate metabolism (particularly diabetes mellitus) the accumulation of ketone bodies adds to acid production.

The large supply of metabolic acids must be neutralized and buffered in order to maintain pH values compatible with life.

SOURCES OF BASES

Biologic systems can also be subjected to excess bases, but these are usually of a dietary or therapeutic nature. Consumption of sodium salts of organic acids leads to the equivalent of NaOH. As an example, when sodium citrate is consumed, the organic portion (citrate) is completely oxidized to CO_2 and H_2O, leaving the Na behind to behave as NaOH.

282   Na citrate $\longrightarrow$ NaOH + CO_2 + H_2O

The NaOH must be neutralized, the same as if an equivalent amount of NaOH had been consumed. Excessive use of sodium bicarbonate can lead to a situation of excess base.

ISOHYDRIC MECHANISM

There must be mechanisms for regulating the acidity within living systems. As previously indicated, CO_2 is produced in large quantities at the tissue level. From the tissues, it enters the red cells (Fig. 14-1).

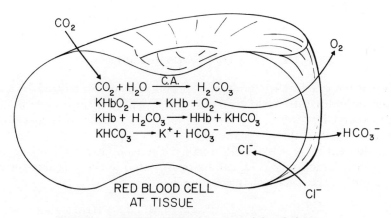

Fig. 14-1. Isohydric mechanism at peripheral tissues.

In the red cell, CO_2 combines with H_2O to produce carbonic acid in a reaction catalyzed by carbonic anhydrase. The low oxygen levels of the tissues promote the dissociation of potassium oxyhemoglobin. Oxyhemoglobin is a stronger acid than hemoglobin, so when it releases oxygen it becomes a weaker acid. The potassium hemoglobin (KHb) reacts with carbonic acid to yield protonated hemoglobin (HHb) and potassium bicarbonate. The red cell is permeable to bicarbonate, and as the concentration of bicarbonate increases it diffuses into the surrounding plasma. Potassium is impermeable to the red cell and is trapped inside. To maintain electrical neutrality within the cell, as bicarbonate diffuses out, chloride, also freely permeable, diffuses in.

At the lungs, the processes are reversed. Protonated hemoglobin combines with oxygen to form protonated oxyhemoglobin, a stronger acid, that reacts with potassium bicarbonate to produce potassium oxyhemoglobin and carbonic acid (Fig. 14-2).

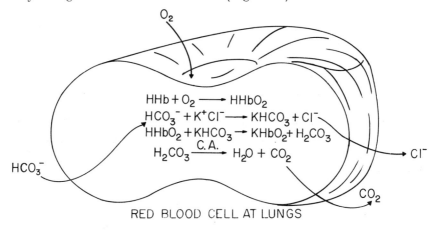

Fig. 14-2. Isohydric mechanism at lungs.

The carbonic acid then undergoes decomposition, catalyzed by carbonic anhydrase to yield water and CO_2. The CO_2 pressure at the lungs (partial pressure) is very low, and CO_2 is able to escape from the red cell into the alveolar spaces and is expired.

This entire process is often called the isohydric mechanism, because large amounts of CO_2 can be transported in the vascular system with only very small changes in pH.

283 $$KHb + H_2CO_3 \rightleftharpoons KHCO_3 + HHb$$

Potassium hemoglobin, a salt of a weak acid, reacts with a rather strong acid, carbonic acid, to produce potassium bicarbonate and protonated hemoglobin, a very weak acid. Therefore, hemoglobin serves as a very effective buffer during the transport of the CO_2 from the tissue to the lungs.

BUFFER SYSTEMS

There are other compounds comprising buffer systems in plasma.

284 $NaHCO_3/H_2CO_3$ Na protein/H protein

         Na_2HPO_4/NaH_2PO_4 Na org acid/H org acid

One of the most important of these buffer pairs is the bicarbonate-carbonic acid buffer. The pK of carbonic acid is 6.1, and at pH 7.4 the bicarbonate/carbonic acid ratio in normal plasma is approximately 20:1.

285
$$pH = pK + \log \frac{[HCO_3]}{[H_2CO_3]}$$

$$7.4 = 6.1 + \log \frac{20}{1}$$

$$7.4 = 6.1 + 1.3$$

pH REGULATION: ROLE OF LUNGS

The importance of the bicarbonate-carbonic acid buffer pair is illustrated by the role of the lungs and the kidneys in the regulation of pH. The bicarbonate-carbonic acid buffer response to acids is the conversion of bicarbonate to carbonic acid, which can be converted to CO_2 and expired at the lungs.

286 $NaHCO_3 + HA \longrightarrow NaA + H_2CO_3$
                                              $\downarrow$
                                         $H_2O + CO_2$

The buffer response to base is the reaction of carbonic acid with a base to produce more bicarbonate and water.

287 $H_2CO_3 + BOH \longrightarrow BHCO_3 + H_2O$

In order to have a pH compatible with life processes, the ratio of bicarbonate to carbonic acid must be approximately 20:1 **(285)**. If the organism is faced with an abundance of acid, such as in diabetes mellitus where there is an accumulation of ketone bodies, **286,** or when more than normal quantities of carbonic acid are produced, the 20:1 ratio can be maintained by expiring more CO_2 at the lungs. This is accomplished by increasing the ventilation rate. Similarly, if there is an excess of basic material whereby the carbonic acid portion is consumed **(287)** resulting in an increased concentration of bicarbonate, the lungs can compensate for this situation by decreased ventilation. Thus, carbonic acid is increased to maintain the proper ratio.

pH REGULATION: ROLE OF KIDNEY

These adjustments in the ratio of bicarbonate to carbonic acid as effected by the lungs are very fast, but the important thing to remember is that the lungs can move only the carbonic acid portion of the buffer pair. The kidneys regulate the bicarbonate. Consider the situation where large quantities of acids are encountered such that the bicarbonate is consumed **(286)**. There is very definitely a limited amount of bicarbonate available to buffer acids.

One of the ways the kidney can regulate the level of bicarbonate in blood is shown in Figure 14-3. This scheme shows the reabsorption

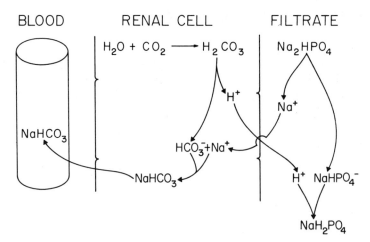

Fig. 14-3. Renal reabsorption of sodium ion and formation of bicarbonate.

of sodium, which combines with bicarbonate to maintain blood bicarbonate levels. Concomitantly, Na_2HPO_4 is converted to NaH_2PO_4, a more acid form of the salt, for excretion.

The kidney can also produce ammonia as a vehicle for proton excretion (Fig. 14-4). Ammonia, released from amino acids in the

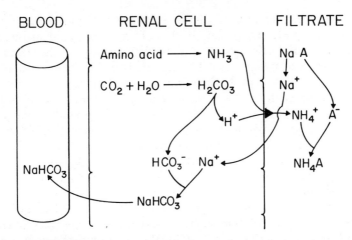

Fig. 14-4. Renal formation of ammonia; secretion of proton as ammonium ion with concomitant reabsorption of sodium ion.

kidneys, combines with a proton from carbonic acid to form the ammonium ion. Sodium is reabsorbed from the filtrate, leaving an anion, A^- ($NaHPO_4^-$, $CH_3\overset{\overset{O}{\|}}{C}-CH_2-COO^-$, etc.), to be excreted as the ammonium salt. The reabsorbed sodium combines with the bicarbonate to maintain blood levels. These mechanisms just described reveal how the kidney is able to conserve sodium bicarbonate and secrete protons. The kidney is effective in this regard as evidenced by the fact that it can acidify urine to a pH of about 5 starting with a filtrate which has a pH of about 7.35.

The kidney is also able to void excessive quantities of bicarbonate that might accumulate from reaction (**287**) or from an overconsumption of bicarbonate, simply by excreting sodium bicarbonate. The sodium bicarbonate in the filtrate is not reabsorbed but allowed to pass with the urine (Fig. 14-3).

From the foregoing it can be concluded that the lungs make fast adjustments in the vascular pH by retention or elimination of carbonic acid. It remains, however, for the kidney to remove the nonvolatile portions of buffer systems such as the bicarbonate, the phosphate, and in some cases the ketone body acid/salt pairs. It should be pointed out that the ability of the kidney to conserve sodium bicarbonate is limited.

BICARBONATE AND CARBONIC ACID MEASUREMENTS

When clinically indicated, the evaluation of patient's acid-base status is important and is usually done by measurement of bicarbonate-carbonic acid buffer system. This system is used rather than some other system because it is easily measured and it includes acid-base regulation as effected by both pulmonary and kidney functions.

A solution such as blood, urine, and spinal fluid containing bicarbonate and carbonic acid can be acidified with an acid stronger than carbonic acid to a pH less than 5, which converts the bicarbonate to carbonic acid. At pH values less than 5 carbonic acid is unstable, decomposing to produce water and CO_2. Since CO_2 is very insoluble at acid pH values, it separates from the solution as a gas.

The volume of gas released can be measured with suitable manometric or volumetric apparatus. One mole of an ideal gas under conditions of standard temperature and pressure occupies a volume of 22.4 liters. Since CO_2 is not an ideal gas, but a real one, it occupies a volume of 22.3 liters (a millimole occupies 22.3 ml) at standard temperature and pressure. Thus, the millimoles of CO_2 can be determined by measurement of CO_2 volume released from a sample upon acidification and subjection to negative pressure. A normal serum or plasma sample will yield approximately 26 mmoles of CO_2 per liter.

In passing, it should be noted that there are colorimetric methods, utilizing the color change of pH indicators, commonly used to measure bicarbonate and carbonic acid in biologic specimens.

Evaluation of acid-base status is incomplete if only the bicarbonate/carbonic acid buffer system is examined. Usually the concentrations of bicarbonate and carbonic acid are about 25 and 1.25 meq/liter, respectively.

288
$$pH = pK + \log \frac{25}{1.25}$$
$$= 6.1 + 1.3 = 7.4$$

Consider the situation where the bicarbonate/carbonic acid ratio is changed from 25:1.25 to 25:1.9. If the carbonic acid concentration is increased to 1.9 meq/liter while the bicarbonate concentration remained the same, an abnormally low pH would result.

289
$$pH = 6.1 + \log \frac{25}{1.9}$$
$$= 6.1 + \log 13.3$$
$$= 6.1 + 1.1 = 7.2$$

The accuracy and precision of bicarbonate and carbonic acid measurements are generally inadequate to detect the differences in the above examples.

pH MEASUREMENTS

A complete evaluation of an acid-base status requires a measurement of pH. Instruments for very accurate and precise measurements of pH are commercially available and are in routine use in most clinical laboratories.

PARTIAL PRESSURE

The concentration of gases is often expressed in terms of partial pressure. For example, the atmosphere contains approximately 80% N_2 and 20% O_2 and has a total atmospheric pressure sufficient to support a column of mercury 760 mm high. The partial pressure can be calculated by multiplying the total pressure (mm Hg) by the fraction concentration of O_2 or N_2 present.

290

$$760 \text{ mm Hg} \times 0.20 \text{ } O_2 = 152 \text{ mm Hg } P_{O_2}$$
$$760 \text{ mm Hg} \times 0.80 \text{ } N_2 = 608 \text{ mm Hg } P_{N_2}$$

OXYGEN TRANSPORT

Partial pressure dictates gaseous exchange at tissue levels and in the lungs. As calculated above, the partial pressure of atmospheric oxygen is approximately 150 mm Hg. In contrast, the partial pressure of O_2 within the lungs is 100 mm Hg and is extremely low at tissue levels owing to the continuous consumption. This decrease in P_{O_2} from atmosphere to tissue promotes the inflow of oxygen. However, the insolubility of O_2 in the aqueous vascular system prohibits diffusion in sufficient quantities. By necessity then, O_2 is transported in combination with hemoglobin.

CARBON DIOXIDE TRANSPORT

As opposed to oxygen, a decrease in CO_2 concentration occurs from tissue to atmosphere, promoting egress of CO_2. The P_{CO_2} in venous return is approximately 45 mm Hg, in the lungs about 40 mm

188   Clinical Biochemistry

Hg, and in the atmosphere 0.5 mm Hg. Like oxygen, CO_2 is relatively insoluble in aqueous systems, so for effective transfer it is transported as bicarbonate. At tissue levels where P_{CO_2} is high, carbonic acid is formed in the erythrocytes by a reaction catalyzed by carbonic anhydrase (Fig. 14-1). In the lungs, the reverse reaction takes place because of low partial pressure of CO_2 (Fig. 14-2). The reaction in both directions is catalyzed by the same enzyme.

291 $$H_2O + CO_2 \xrightleftharpoons[\text{anhydrase}]{\text{carbonic}} H_2CO_3$$

The difference between venous and atmospheric P_{CO_2} has practical significance in the laboratory. If a blood sample has a P_{CO_2} of 40 mm Hg and is exposed to the atmosphere during centrifugation or sample allocation, it will rapidly lose CO_2 into the atmosphere. Carbonic acid is the source of this CO_2.

292 $$H_2CO_3 \rightleftharpoons H_2O + CO_2 \uparrow$$

Prior to analysis, a blood sample exposed to the atmosphere for an hour or so will have a carbonic acid content that is practically nil and a pH of about 8. If the CO_2 produced upon acidification of this sample is measured, manometrically or colorimetrically, it will be mostly bicarbonate. The carbonic acid lost is small since it represents 1/20 of the total CO_2. For normal levels of bicarbonate, 1/20 is equal to 1.25 meq/liter.

293 $$25 \text{ meq/liter} \times \frac{1}{20} = 1.25 \text{ meq/liter}$$

METHODS FOR EVALUATING ACID-BASE BALANCE

Reaction **292** provides the basis for two very good methods for evaluating acid-base balance. The reaction shows that the concentration of carbonic acid depends upon the concentration of CO_2. The concentration of the latter depends upon the P_{CO_2} of the particular environment. As stated above, if a blood sample is exposed to the atmosphere with a very low P_{CO_2}, most of the CO_2 and H_2CO_3 is lost from the sample. At the tissues, where the P_{CO_2} is relatively high, the reverse reaction with the formation of H_2CO_3 occurs. Both of the methods utilize the principle stated in the Henderson-Hasselbalch equation.

294 $$pH = pK + \log \frac{[HCO_3^-]}{[H_2CO_3]}$$

As indicated above, the H_2CO_3 concentration is directly proportional to the Pco_2. This can be written as an equality.

295 $$[H_2CO_3] = Pco_2 \times 0.03$$

Substituting into **294** gives

296 $$pH = pK + \log \frac{[HCO_3^-]}{[Pco_2 \times 0.03]}$$

If equation **296** is rewritten **297** is obtained.

297 $$pH = pK + \log [HCO_3^-] - \log Pco_2 - \log 0.03$$

The values of pK (6.1) and 0.03 are constant. For any given sample the $[HCO_3^-]$ is also constant. The sum of constants is just another constant.

298 $$-\log 0.03 + pK + \log [HCO_3^-] = K$$

and substituting into **297** provides

299 $$pH = K - \log Pco_2$$

In this form it is easily seen that an equation for a straight line with a negative slope is obtained. A plot would appear as shown in Figure 14-5.

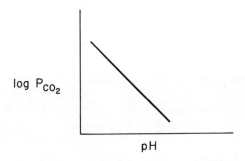

Fig. 14-5. Relationship of Pco_2 and pH as described in Equation 299.

Pco_2 DETERMINATION: EQUILIBRIUM METHOD

In the equilibration method, the pH of the blood sample is measured following collection and maintenance under anaerobic conditions. Next, an aliquot of the same sample is equilibrated with an oxygen–CO_2 atmosphere with a P_{CO_2} of 60 mm Hg. When equilibration is complete, the pH of the sample is measured. Finally, another aliquot is equilibrated with a gas mixture with a low P_{CO_2}, 30 mm Hg. Following this the third pH measurement is made. In the equilibration technique the blood sample is equilibrated with 2 different atmospheres with known values of P_{CO_2}, and the resulting pH values are measured. These points are plotted on a graph as in Figure 14-6 as points *A* and *B*.

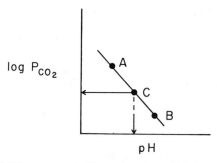

Fig. 14-6. Measurement of P_{CO_2} by the equilibrium method.

A straight line drawn through these two points then describes the relationship of P_{CO_2} and pH for that sample. The initial pH value for the unmanipulated sample is plotted on the line, point *C*, and the original P_{CO_2} for the sample is found by drawing a line to the ordinate parallel to the abscissa. From these measurements the original pH and P_{CO_2} of the sample are obtained. By the use of **296** the $[HCO_3^-]$ can be calculated.

P_{CO_2} DETERMINATION: DIRECT METHOD

The second technique is called the P_{CO_2} electrode method. The equipment required has a sample chamber separated from a dilute bicarbonate solution by a membrane permeable to CO_2 (Fig. 14-7).

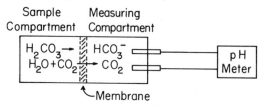

Fig. 14-7. Schematic representation of a device for the direct measurement of P_{CO_2}.

As seen in equation **296,** the P_{CO_2} is indirectly proportional to the pH. Thus, when a sample containing H_2CO_3 or CO_2 is placed in the sample compartment CO_2 will diffuse into the measuring compartment and lower the pH. A sample with a high P_{CO_2} will provide more CO_2 for the measuring compartment and thus result in a lower pH than a sample with a low P_{CO_2}. Calibration of the pH meter by the use of samples with known P_{CO_2} allows a readout of P_{CO_2}.

CHAPTER 15

ELECTROLYTES

The preceding chapters have dealt mainly with consideration of organic compounds; however, electrolytes are also essential constituents in biochemical processes. They are important in the maintenance of water balance and distribution, in osmotic equilibrium, and in regulating neuromuscular irritability. Their role in acid-base balance was discussed previously.

Consider the mammalian organism as consisting of several fluid compartments, the two major ones being the intracellular and extracellular. The latter can be subdivided into interstitial fluid and blood plasma. In these various compartments the electrolyte concentrations differ, but they are in equilibrium. Sampling problems limit their measurement to plasma or serum. The main electrolytes and normal concentrations are listed in Table 15-1.

Table 15-1. SERUM ELECTROLYTE CONCENTRATIONS

Electrolyte	Concentration
Cations	
Sodium (Na^+)	136.0 – 144.0 meq/liter
Potassium (K^+)	3.8 – 5.1 meq/liter
Calcium (Ca^{++})	4.3 – 5.2 meq/liter
Magnesium (Mg^{++})	1.5 – 2.4 meq/liter
Anions	
Chloride (Cl^-)	95.0 – 105.0 meq/liter
Bicarbonate (HCO_3^-)	24.0 – 27.0 meq/liter
Phosphate (P_i)	2.5 – 4.5 mg/dl

These elements must be considered as essential for the welfare of the organism, and because of their continuous loss in urine, feces, and sweat, they must be replaced in the diet. Since all except calcium are very common in ordinary foodstuffs, there are seldom circumstances where there are inadequate supplies. Indeed, it is difficult to have a diet low in NaCl, commonly referred to as a low salt or " salt-free " diet.

The electrolytes sodium, potassium, and chloride are of concern because they are most apt to show changes in disease. The concentrations of sodium and potassium are directly proportional to neuromuscular irritability (function). This is illustrated by the fact that potassium concentrations of 9 meq/liter or more are usually fatal. Likewise, potassium concentrations less than 2 meq/liter can also be fatal.

Deficiencies of sodium, potassium, and chloride result from having an increased loss or a decreased intake. Less than normal serum concentrations of potassium, sodium, and chloride are referred to as hypokalemia, hyponatremia, and hypochloremia, respectively. Losses can occur by several routes: (1) the gastrointestinal tract – either in the form of vomiting or diarrhea, as a result of therapy, and occasionally by excessive aspiration of gastric or intestinal contents; (2) the skin – large losses can occur in the form of excessive sweating; and (3) the kidney – urine is the primary route of excretion; but normally the kidney is very effective in maintaining adequate serum levels but in some diseases excessive losses occur which result in deficiencies.

Elevations above normal serum concentrations of sodium, potassium, and chloride (hypernatremia, hyperkalemia, and hyperchloremia) can occur as a result of dehydration. This type of water loss happens without the concomitant loss of electrolytes or by inadequate water intake. However, sodium, potassium, and chloride are usually depleted along with water losses. Improper electrolyte replacement therapy can also cause elevated concentrations.

The electrolyte concentrations in urine are extremely variable due to differences in volume and diet. Under conditions of known dietary intake the amounts of sodium and potassium excreted and their concentration ratio can be clinically useful.

METHODS FOR MEASURING ELECTROLYTES

SODIUM AND POTASSIUM

The measurement of sodium and potassium in biologic materials is best done using either flame emission or flame absorption.

CHLORIDE

Chloride analysis is usually accomplished by titrimetric or colorimetric methods. In a common titrimetric procedure a standard or an unknown sample containing chloride is titrated with a $Hg(NO_3)_2$ solution in the presence of an indicator (IND) containing *sym*-diphenylcarbazone. The mercuric ion combines avidly with chloride forming mercuric chloride. When all the chloride present has reacted, an excess of mercuric ions accumulates in the solution. The accumulated mercuric ions combine with the indicator to form a purple complex.

300
$$Hg(NO_3)_2 + 2\,Cl^- \longrightarrow HgCl_2 + 2\,NO_3^-$$
$$Hg(NO_3)_2 + IND \longrightarrow Hg \cdot IND + 2\,NO_3^-$$
$$\text{colorless} \qquad\qquad \text{purple}$$

The chloride concentration of a sample also can be determined colorimetrically by reaction with mercuric thiocyanate.

301
$$Hg(SCN)_2 + 2\,Cl^- \longrightarrow HgCl_2 + 2\,SCN^-$$
$$3\,SCN^- + Fe(NO_3)_3 \longrightarrow Fe(SCN)_3 + 3\,NO_3^-$$
$$\text{red}$$

As in the titration technique described, the mercuric ion reacts with chloride to form weakly ionizable $HgCl_2$. The thiocyanate then reacts with the iron ions to produce red ferric thiocyanate. This color is directly proportional to the amount of chloride present.

CALCIUM

As an electrolyte, calcium is concerned with the clotting of blood, cell permeability, and is a cofactor for enzymes. It is extremely important in normal neuromuscular irritability, and its effects are indirectly proportional to its concentration. Low levels of calcium lead to hyperneuromuscular irritability called tetany.

Concentrations of calcium, as usually measured in serum, represent the total calcium. However, by suitable techniques it is possible to show that approximately one-half the calcium in serum is ionic and one-half is tightly bound to proteins and does not behave as free calcium ions. Only the free calcium is physiologically active.

The classic method of calcium analysis in biologic material is the precipitation of the calcium from solution by oxalate to form very insoluble calcium oxalate.

302

$$Ca^{++} + \begin{array}{c} O \\ \parallel \\ C-ONa \\ | \\ C-ONa \\ \parallel \\ O \end{array} \longrightarrow \begin{array}{c} O \\ \parallel \\ C-O \\ | \quad \diagdown Ca \\ C-O \diagup \\ \parallel \\ O \end{array} + 2\,Na^+$$

$$\begin{array}{c} O \\ \parallel \\ C-O \\ | \quad \diagdown Ca \\ C-O \diagup \\ \parallel \\ O \end{array} + H_2SO_4 \longrightarrow CaSO_4 + \begin{array}{c} COOH \\ | \\ COOH \end{array}$$

$$5\,\begin{array}{c} COOH \\ | \\ COOH \end{array} + 2\,KMnO_4 + 3\,H_2SO_4 \longrightarrow$$

$$K_2SO_4 + 2\,MnSO_4 + 8\,H_2O + 10\,CO_2$$

The precipitate is washed to remove other materials, dissolved in H_2SO_4, and titrated with potassium permanganate. When the solution is hot the permanganate oxidizes the oxalic acid. After the oxalic acid has been completely oxidized, the addition of more permanganate forms a persistent purple permanganate color, indicating the titration end point.

The estimation of calcium utilizing the oxalate-permanganate method is primarily of historic interest. It has little application in modern laboratories. The difficulty with this technique is that a reasonable amount of serum, on the order of 0.5 ml, contains approximately 50 µg of calcium. To collect this amount of precipitate and wash it free of oxalic acid and other materials which react with the permanganate, without losing the desired calcium oxalate, is difficult.

A better way of measuring calcium is by chelometric titration using metal-sensitive indicators.

To a solution containing calcium, a metal-sensitive indicator is added which combines with calcium to form a calcium-indicator complex. Initially, the indicator is one color, color x, but when combined with calcium a different color is formed, color y. After the calcium-indicator complex is formed, it is titrated with an EDTA solution. EDTA forms a complex with the calcium called a chelate **(303** (a), (b), (c)**)**.

303

(a) $Ca^{++} + Ind^= \longrightarrow Ca \cdot Ind$
    color x color y

(b) $Ca \cdot Ind + EDTA \longrightarrow Ca \cdot EDTA + Ind^=$
    color y color x

(c)
$$\underset{\substack{\text{disodium} \\ \text{ethylenediaminetetraacetate} \\ \text{(EDTA)}}}{\begin{array}{c} COOH \\ | \\ CH_2 \\ | \\ N-CH_2-CH_2-N \\ | \hspace{2.2cm} | \\ CH_2 \hspace{1.5cm} CH_2 \\ | \hspace{2.2cm} | \\ COONa \hspace{0.8cm} COONa \end{array}} + Ca^{++} \longrightarrow \underset{\text{calcium chelate}}{\begin{array}{c} COOH \hspace{2cm} COOH \\ | \hspace{3.5cm} | \\ CH_2 \hspace{2cm} CH_2 \\ | \hspace{3.5cm} | \\ N-CH_2-CH_2-N \\ | \hspace{1cm} \diagdown \hspace{0.5cm} \diagup \hspace{1cm} | \\ CH_2 \hspace{0.5cm} Ca^{++} \hspace{0.5cm} CH_2 \\ | \hspace{1cm} \diagup \hspace{0.5cm} \diagdown \hspace{1cm} | \\ O^- \hspace{1.5cm} O^- \\ | \hspace{3.5cm} | \\ C \hspace{2cm} C \\ \| \hspace{3.5cm} \| \\ O \hspace{2cm} O \end{array}} + 2\,Na^+$$

A chelate is a complex structure which differs from a salt in that the ion chelated remains ionic but firmly bound to the chelating substance. The EDTA binds calcium (**303** (b)) much more strongly than the indicator and is able to kidnap the calcium ion, producing a color change—that is, it changes from color y to x. Thus, when color x is restored, the end point has been reached. For quantitation, the unknown titer is compared with the titer obtained for known aqueous standards titrated under identical conditions.

For the chelometric assay of calcium the reaction medium is alkaline to prevent interference by magnesium. Magnesium forms the unreactive hydroxide, which effectively removes it from the reaction.

304 $\hspace{2cm} Mg^{++} + 2\,OH^- \longrightarrow Mg(OH)_2$

The metal-sensitive indicators react like pH indicators.

305 $\hspace{2cm} \underset{\text{colorless}}{HInd} + OH^- \longrightarrow \underset{\text{colored}}{Ind^-} + H_2O$

In its acid form a pH indicator such as phenolphthalein is colorless, but when the proton is removed by a suitable base it forms a colored anion. Similarly, a calcium-sensitive indicator will have one color in the absence of calcium and another color when it is present.

A number of metal-sensitive indicators are available. Some are more applicable to clinical use than others. There is at least one metal-sensitive indicator which shows fluorescence in the absence of calcium, and a change in fluorescence indicates the end point.

The difficulty with the chelometric titrations employing metal-sensitive indicators is the detection of the end point. There are two colors involved, y at the beginning and x at the end (**303** (b)). If colors y and x are very distinctive, the end point is relatively easy to detect, but if the colors are similar, the equivalence point is difficult to determine. Also, during the titration the solution will contain a mixture of both colors, and for some indicators the complete absence of color y is difficult to discern.

The best technique by far for assaying calcium is atomic absorption spectrophotometry.

MAGNESIUM

Magnesium behaves in many respects like calcium: it is required by many enzymes and its effects in neuromuscular irritability are indirectly related to concentration. Low levels of magnesium can also lead to tetany. This is sometimes seen in severe malnutrition, particularly in alcoholics. High concentrations of magnesium have the expected opposite effects, and indeed it is possible to anesthetize animals with very high magnesium doses.

Magnesium may be measured by chelometric titrations employing metal-sensitive indicators or by colorimetric techniques. Just as it is true for calcium, however, the best technique for measuring magnesium is atomic absorption spectrophotometry.

PHOSPHATE

The importance of phosphorus is obvious from the numerous previous encounters with this substance, both as a buffer system and as a component of the various organic molecules. Phosphorus deficiencies are extremely rare because phosphorus is so ubiquitous. A significant aspect of phosphorus metabolism arises from the fact that it is intimately related to calcium levels.

306 $$(\text{mg/dl Ca}) \times (\text{mg/dl Pi}) \approx 40$$

The product of the concentrations of calcium and phosphorus equals a relatively constant number. From this it would be expected that increases in serum calcium levels would lead to decreased levels of phosphorus and vice versa, and indeed that is usually so.

Phosphorus is usually measured by colorimetric techniques, and they are all very similar. The sample containing phosphate is acidified with a stronger acid to form phosphoric acid, which reacts with molybdic acid to form phosphomolybdic acid (**307** (a)). Upon reduction, phosphomolybdic acid yields molybdenum blue, a colloidal material the color of which is proportional to the phosphorus present (**307** (b)).

307 (a) $H_3PO_4 + 12\ H_2MoO_4 \longrightarrow H_3PO_4 \cdot 12\ MoO_3 + 12\ H_2O$

(b) $H_3PO_4 \cdot 12\ MoO_3 + 8\ e + 8\ H^+ \longrightarrow 4\ (MoO_2)_2MoO_4 + H_3PO_4 + 4\ H_2O$

molybdenum blue

The various methods for measuring phosphorus vary primarily in the reducing agents used to convert phosphomolybdic acid to the molybdenum blue. The following are some of the commonly used reducing agents.

308 $\qquad 4\ Sn^{++} \longrightarrow 4\ Sn^{++++} + 8\ e$

309

$4\ CH_3-\underset{H}{N}-\!\!\!\left\langle\ \right\rangle\!\!\!-OH \longrightarrow 4\ CH_3-N=\!\!\!\left\langle\ \right\rangle\!\!\!=O + 8\ H^+ + 8\ e$

p-methylaminophenol

310

[1-amino-2-naphthol-4-sulfonic acid reacts to form the corresponding imino-quinone] $+ 8\ H^+ + 8\ e$

1-amino-2-naphthol-4-sulfonic acid

311 $\qquad 8\ NH_2OH \longrightarrow 4\ N_2 + 8\ H_2O + 8\ H^+ + 8\ e$

hydroxylamine

In addition, several other reagents have enjoyed some popularity.

ELECTROLYTE INTERRELATIONSHIPS

The above discussion of electrolytes is a micro-mini coverage of the subject at best. The subject is quite complex because of the interrelationships of most of the electrolytes. For instance, neuromuscular function is directly related to the sum of the sodium and potassium concentrations and inversely related to the sum of calcium and magnesium and proton concentrations.

312
$$\text{irritability} \propto \frac{[Na^+] + [K^+]}{[Ca^{++}] + [Mg^{++}] + [H^+]}$$

Thus, normal muscular activity is dependent not only on the electrolytes discussed but also upon the acid-base balance. Also, the secretion and in some cases the absorption of electrolytes is regulated by hormones. Sodium and potassium secretion is largely regulated by hormones produced in the adrenal cortex, the so-called mineralcorticosterods. Calcium and phosphorus metabolism is made more complex since it is also affected by vitamin D. To elucidate further, the serum concentrations of calcium and phosphorus are elevated above normal adult levels in children, which is related to the formation of bone. The sheer complexity of this subject make further discussion beyond the scope of this book.

CHAPTER 16

HORMONES

The endocrine system consists of glands without ducts that produce regulatory substances called hormones. The hormones range from simple to rather large, complex protein molecules. The glandular secretions are released directly into the vascular system, and the hormones usually have their effects on tissues or organs remote from the source. Since most hormones are extremely potent, only small amounts are required to produce a dramatic effect.

ANTERIOR PITUITARY

The pituitary gland, located at the base of the brain, is comprised of two functional parts: anterior and posterior lobes. The anterior lobe secretes tropic hormones and growth hormone all of which are believed to be proteins. The function of these hormones is to stimulate the function or secretion of other endocrine glands. The anterior pituitary elaborates a thyroid stimulating hormone (TSH) which stimulates the thyroid gland to produce thyroxine. Thyroxine affects the secretion of TSH by a negative feedback system. That is, when the circulating level of thyroxine is great enough, the pituitary stops producing TSH. Conversely, when thyroxine in the blood falls below certain levels, the pituitary gland elaborates TSH in order to reestablish appropriate thyroxine levels. This is represented diagrammatically in **313.**

313

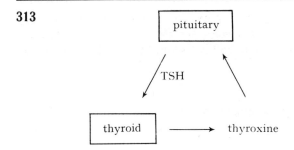

The anterior pituitary is often referred to as the "master gland" because of its crucial role in the regulation of other endocrine organs. The hormones of the anterior pituitary and their abbreviations are listed in Table 16-1.

TABLE 16-1. HORMONES FROM THE ANTERIOR PITUITARY

Name	Abbreviation
Thyroid stimulating hormone	TSH
Growth (somatotropic) hormone	GH
Adrenocorticotropic hormone	ACTH
Prolactin, lactogenic or luteotropic hormone	LTH
Luteinizing hormone or	LH
interstitial cell stimulating hormone	ICSH
Follicle stimulating hormone	FSH

GROWTH HORMONE

Growth hormone (GH), elaborated by the anterior pituitary, regulates growth and development. Excessive production of the hormone during the period of skeletal growth results in gigantism; such individuals are excessively tall, but generally well proportioned. Most of the giants in carnivals are individuals with this condition. If overproduction occurs in adults – that is, after the long bone epiphyses have closed – additional growth results in excessive production of fibrous tissue. This condition is called acromegaly and is characterized by enlarged head, hands, and feet. If the disease is caused by a tumor in the anterior pituitary surgical removal is sometimes successful.

Deficiency of growth hormone during infancy or adolescence can cause dwarfism. The patient, unlike the cretin, is well proportioned and suffers no mental defects. Attempts have been made to treat pituitary dwarfism with growth hormone isolated from pituitary glands. However, because of species specificity, the most effective hormone is obtained from human cadavers and the supply is obviously limited.

POSTERIOR PITUITARY HORMONES

ANTIDIURETIC HORMONE

The posterior pituitary elaborates two hormones: antidiuretic hormone (ADH or vasopressin) and oxytocin (pitocin). ADH controls the removal of water by the kidney. Apparently the osmotic pressure of the blood regulates ADH secretion. When the vascular system becomes too concentrated, osmotic receptors stimulate the production of ADH, preventing or restricting the loss of water by the kidney. If the vascular system tends to be too dilute, less hormone is produced thereby allowing additional water loss via the kidneys. Insufficient quantities of ADH produce the syndrome diabetes insipidus. Individuals with the disease consume large volumes of fluids, perhaps up to 30 liters of water per day, and excrete comparable amounts of urine. Posterior pituitary extracts are used with some success for treatment.

OXYTOCIN

Oxytocin stimulates the contraction of smooth muscles and is often used in obstetrics to induce labor. Both ADH and oxytocin are small peptides each containing 8 amino acids; they both have been synthesized in the laboratory. The properties and characteristics of the synthetic hormones are the same as those of the naturally occurring compounds.

THYROID

The thyroid gland is located in the neck. It is able to convert tyrosine, one of the essential amino acids, and inorganic iodine into the hormone thyroxine (T_4).

314

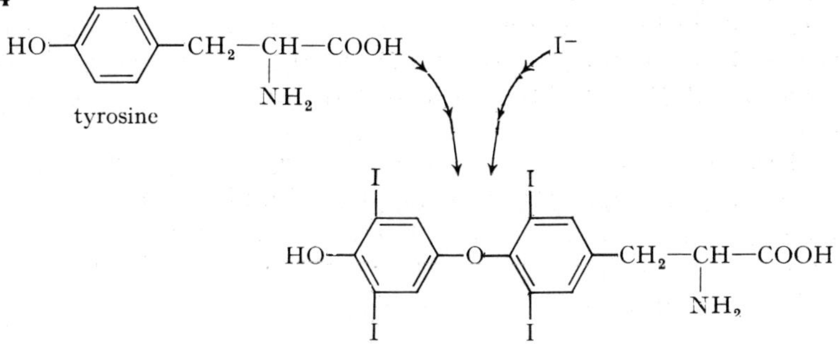

$HO-\langle\rangle-CH_2-CH(NH_2)-COOH$
tyrosine

I^-

thyroxine
(3,5,3',5'-tetraiodothyronine)
T_4

$HO-\underset{I}{\overset{I}{\langle\rangle}}-O-\underset{I}{\overset{I}{\langle\rangle}}-CH_2-CH(NH_2)-COOH$

T_4 is a common abbreviation for thyroxine, which is thyronine with 4 iodine atoms attached. Another compound containing 3 atoms of iodine, T_3, has also been isolated from the thyroid gland in small amounts, and is in some situations more active than thyroxine.

315

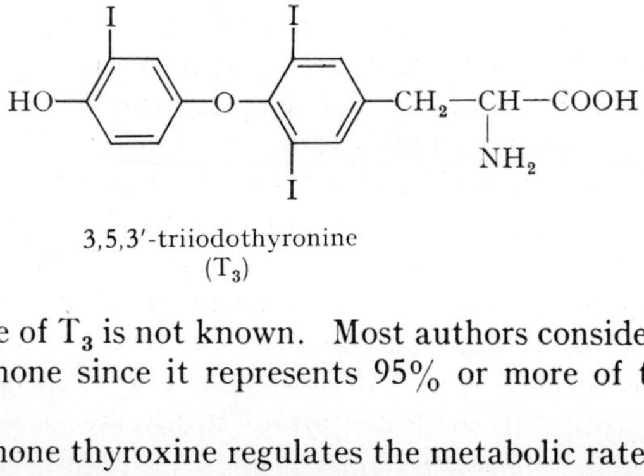

$HO-\overset{I}{\langle\rangle}-O-\underset{I}{\overset{I}{\langle\rangle}}-CH_2-CH(NH_2)-COOH$

3,5,3'-triiodothyronine
(T_3)

The exact role of T_3 is not known. Most authors consider T_4 to be the primary hormone since it represents 95% or more of the output of the thyroid.

The hormone thyroxine regulates the metabolic rate in mammals. However, the exact site and mechanism of action of T_3 and T_4 have not been elucidated. Decreased amounts of thyroxine result in a reduced metabolic rate. Excessive amounts of thyroxine produce the opposite effect.

HYPOTHYROIDISM

Effects of hypo- and hypersecretion of the thyroid gland cause several diseases of clinical importance. Insufficient quantities of thyroxine – hypothyroidism – can occur as a result of inadequate stimulation of the gland by the anterior pituitary via TSH, or as a result of primary disease of the gland. If this condition occurs in utero it leads to cretinism. The cretin is a disproportionate dwarf and is mentally retarded. Many of the adverse effects can be reversed if the condition is detected and treated at birth with thyroxine.

Hypothyroidism, particularly among adults, is a rather common condition and is frequently treated by the administration of desiccated thyroid gland from domesticated animals or thyroxine. Because thyroxine is a small molecule and is not subject to digestion, it can be administered orally.

HYPERTHYROIDISM

Hyperthyroidism can be caused by hyperplasia or by neoplasm of the thyroid gland. The condition is marked by increased metabolic rate, hypertension, tachycardia, extreme nervousness, and apprehension. Thyroid hyperplasia can also be manifest as exophthalmic goiter, which is characterized by protrusion of the eyeballs.

There are several methods of treatment available for hyperthyroidism. One is the administration of Lugol's iodine solution (aqueous potassium iodide and iodine). Small quantities of iodide are stimulatory, but large amounts inhibit thyroxine release. Hyperthyroidism due to hyperplasia or malignancies can be treated with radioactive iodine (^{131}I). This material is concentrated in the thyroid gland and furnishes lethal radiation. By destroying some of the gland, thyroxine production is decreased. Also, part of the gland can be removed surgically. The difficulty with these latter techniques is limiting the destruction or surgical removal to only the appropriate amount of the gland necessary to achieve normal thyroxine secretion. Drugs, such as thiourea, thiouracil, and 2-thio-6-propyluracil, can also be effective in treating hyperthyroidism.

316

thiourea   2-thiouracil   2-thio-6-propyluracil

Hyperplasia of the thyroid gland due to a lack of inorganic iodine in the diet causes simple goiter. This can usually be corrected by simply incorporating sodium iodide in the diet.

ASSAYS OF THYROID HORMONE

The quantity of thyroxine and T_3 in the vascular system is small and more than 99.5% is reversibly bound to plasma proteins the major one being thyroxine binding globulin (TBG). The bound hormones are believed to be physiologically inactive. It is the free hormone which enters the cells and evokes the physiologic response. However, even though the bound hormones are inactive, the status of thyroid function can be estimated by a laboratory procedure that measures the amount of protein-bound thyroxine. This procedure is commonly referred to as protein-bound iodine (PBI).

To determine the PBI content of a sample, the iodine not associated with protein, mostly inorganic iodine, must be removed. This can be accomplished by precipitating the proteins and washing the insoluble precipitate or by treating the sample with an insoluble ion-exchange resin that combines with the inorganic iodide. The bound iodine is released from the organic molecules as inorganic iodine by destructive digestion. Quantitation of the released iodine is accomplished by reaction of sulfatoceric and arsenious acids, catalyzed by inorganic iodine.

317

$$2\ H_4Ce(SO_4)_4 + H_3AsO_3 + H_2O \xrightarrow{I^-}$$

sulfatoceric acid   arsenious acid

$$2\ H_3Ce(SO_4)_3 + H_3AsO_4 + 2\ H_2SO_4$$

sulfatocerous acid   arsenic acid

This reaction is very slow in the absence of inorganic iodine. In the presence of inorganic iodine the reaction rate is greatly increased and is proportional to the iodine concentration. Sulfatoceric acid is yellow-orange, whereas sulfatocerous acid is colorless, and the decrease in the sulfatoceric acid color is a measure of the extent of the reaction and, in turn, a measure of the inorganic iodine concentration.

The concentration of iodine bound to protein in normal serum is 4 to 8 µg/dl. Since 0.5 or 1.0 ml of serum is usually used for PBI determinations, a value of 6 µg/dl represents only 30 to 60 ng of iodine.

The whole procedure is subject to many sources of contamination due to the minute amounts of iodine involved. The common use of iodide and iodine in the laboratory increases the possibility of glassware and equipment contamination. Therefore, the area in the laboratory utilized for PBI determination is quite often isolated. Various drugs and diagnostic compounds administered to patients can also be a source of contamination. For instance, many radiopaque dyes contain iodine. These compounds often behave exactly like thyroxine in the assay and are measured like PBI, circumstances which lead to erroneous and often elevated values.

The measurement of PBI provides an overall indication of thyroid function, but in some cases it can be misleading. The measurement of unbound hormone ("free thyroxine") correlates more closely with clinical observation. If the level of TBG is high, then the PBI will also be high, but the concentration of unbound thyroxine can be normal. Similarly if the TBG concentration is low, PBI will also be low, but the patient may well be euthyroid.

ADRENAL CORTEX

The pituitary controls the adrenal cortex through adrenocorticotropic hormone (ACTH). The adrenal cortex elaborates more than forty different steroid hormones. These particular steroids are sometimes referred to as corticosteroids. The primary circulating corticosteroid is cortisol or hydrocortisone.

318

cortisol

(hydrocortisone; Δ^4-pregnene-11 β-17 α, 21-triol-3,20-dione)

The cortical hormones have many effects, the primary ones being on mineral and carbohydrate metabolism. They affect sodium and potassium retention or secretion, the conversion of glycogen to glucose (glycogenolysis), and the conversion of amino acids to glucose (gluconeogenesis).

Hyperfunction of the adrenal cortex (Cushing's syndrome) with the production of excessive amounts of the steroid hormones is characterized by a retention of sodium and a loss of potassium in the urine. Also patients with this disease are apt to show many of the features of diabetes mellitus because of the carbohydrate effects of the corticosteroids. The corticosteroids behave as though they antagonize the action of insulin.

The opposite effects are observed in adrenocortical hypofunction. The cause of this hypofunction can be either atrophy of the gland itself (Addison's disease) or inadequate stimulation by the anterior pituitary (Simmond's disease). The two can be differentiated by the administration of ACTH. If the signs and symptoms of hypocortical function are eliminated by the administration of ACTH, then the cause is the lack of stimulation by the pituitary. If, however, there is no response, then the defect lies in the adrenal cortex itself. In Addison's and Simmond's disease, the effect on carbohydrate metabolism is very similar to that of hyperinsulinism. The effect on electrolytes is a loss of sodium in the urine and a tendency to retain potassium.

Cortisol can be measured in blood by a fluorometric technique or it and other corticosteroid endproducts can be measured in the urine by methods which measure 17-hydroxycorticosteroids (refer to **68** for the steroid numbering system).

ADRENAL MEDULLA

The adrenal gland consists of two parts: the adrenal cortex mentioned previously and the adrenal medulla. The adrenal medulla secretes the hormones norepinephrine and epinephrine, which are derived from tyrosine. As indicated earlier, these two substances are sometimes called catecholamines. Secretion of these hormones is under control of the nervous system. Their function is to increase the concentration of blood sugar largely by glycogenolysis, and to some extent, by gluconeogenesis. They also increase the heart rate and therefore raise the blood pressure.

The concept that norepinephrine is involved in "fight" situations whereas epinephrine is concerned with "flight" has been developed from observations. It has been found that professional athletes participating in contact sports such as football and boxing secrete more norepinephrine than epinephrine. Athletes involved in non-contact sports (basketball, baseball, tennis) secrete more epinephrine. Indeed, an individual's personality may well be affected by the production of norepinephrine and epinephrine.

Further metabolism of epinephrine and norepinephrine leads to metanephrine and normetanephrine, respectively, and finally to 3-methoxy-4-hydroxymandelic acid.

319

$$\text{tyrosine (HO-C}_6\text{H}_4\text{-CH}_2\text{-CH(NH}_2\text{)-COOH)} \longrightarrow \text{norepinephrine (3,4-dihydroxyphenyl-CH(OH)-CH}_2\text{NH}_2\text{)}$$

From tyrosine the pathway proceeds to:

- **norepinephrine**: HO–C₆H₃(OH)–CH(OH)–CH₂NH₂
- **normetanephrine**: HO–C₆H₃(OCH₃)–CH(OH)–CH₂NH₂
- **epinephrine**: HO–C₆H₃(OH)–CH(OH)–CH₂NH–CH₃
- **metanephrine**: HO–C₆H₃(OCH₃)–CH(OH)–CH₂NH–CH₃
- **3-methoxy-4-hydroxymandelic acid**: HO–C₆H₃(OCH₃)–CH(OH)–COOH

All five of these compounds are normally excreted in urine. Fluorometric techniques can be used to measure catecholamines, which are normally excreted at rates of 25 to 125 µg/day. After conversion to vanillin, 3-methoxy-4-hydroxymandelic acid is conveniently measured spectrophotometrically.

Normal excretion of 3-methoxy-4-hydroxymandelic acid is 2 to 8 mg/24 hours. It can be measured with greater accuracy and precision than the catecholamines because it occurs in considerably greater amounts. Also, the spectrophotometric method is less subject to interferences than are the fluorometric procedures.

In suspected cases of pheochromocytoma, an assessment of the production of the catecholamines (epinephrine and norepinephrine) and 3-methoxy-4-hydroxymandelic acid is possible by the measurement of the substances in urine. Pheochromocytoma results from hyperplasia or malignancy of the medullary tissue of the adrenal gland or similar cells located in various parts of the body. Characteristic of this neoplasm is the production of excessive quantities of epinephrine and norepinephrine which leads to elevated blood glucose, hypertension, and tachycardia. Another tumor associated with the adrenomedullary system is the malignant neuroblastoma of children. This tumor, as well as pheochromocytomas, may yield elevated levels of catecholamines, metanephrines, and 3-methoxy-4-hydroxymandelic acid.

SEX GLANDS

ESTROGENS

The ovaries produce estrogens in response to the gonadotropic hormones secreted by the anterior pituitary.

320

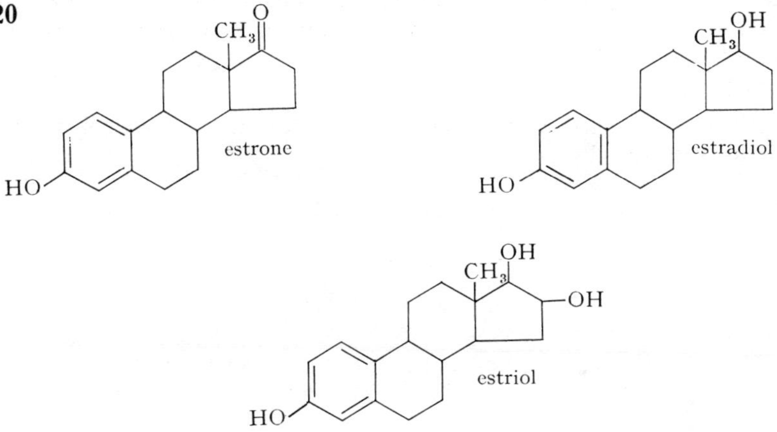

estrone

estradiol

estriol

It is thought that estrone is produced first and is converted to estradiol, which is in turn converted to estratriol or estriol. Their importance is in the development and maintenance of secondary sexual characteristics. In addition they are important in the uterine changes that follow menstruation and the maturation of the graafian follicles, which result in the release of ova. The synthesis of estrogens also takes place in the male testis, but the amount produced is much less than in the female. The measurement of urinary estrogens is of value in the diagnosis of fetal distress during pregnancy.

PROGESTATIONAL HORMONES

Progesterone, which is produced by the ovaries, prepares the uterus for embryonic attachment and inhibits ovulation during pregnancy.

321

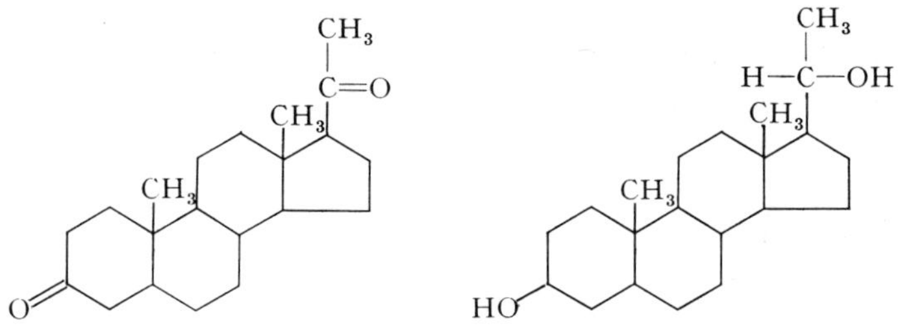

progesterone   *pregnanediol*

Pregnanediol, the inactive metabolic product of progesterone, increases as pregnancy progresses and is excreted in the urine. Low levels of pregnanediol excretion during pregnancy indicate placental dysfunction, which may result in spontaneous abortion or fetal death. Elevations are suggestive of ovarian or adrenocortical tumors.

Oral contraceptives contain purely synthetic steroid compounds such as mestranol.

322

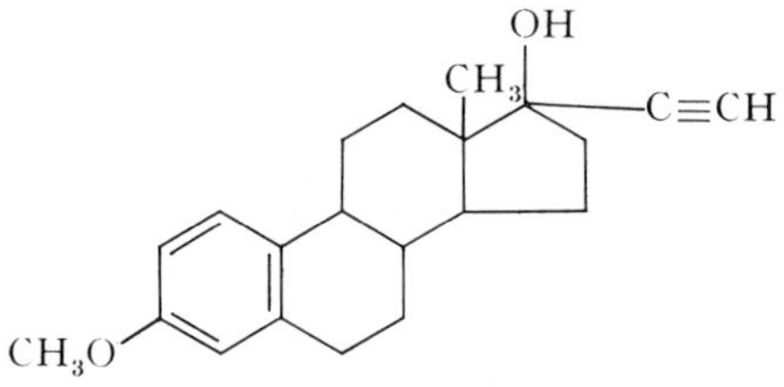

mestranol

They circulate in the vascular system and function in the same way that estrogens would in the feedback system. The anterior pituitary is fooled, so to speak, into believing that sufficient estrogens are circulating and the release of gonadotropic hormones are restricted so ovulation does not take place. Compare the structures of estrogens (**320**) and mestranol (**322**).

ANDROGENS

A predominance of androgens is characteristic of the male. Testosterone is an androgen produced primarily in the testis, but like the estrogens it is not exclusive for one sex. In the female, testosterone is produced by the adrenal cortex.

323

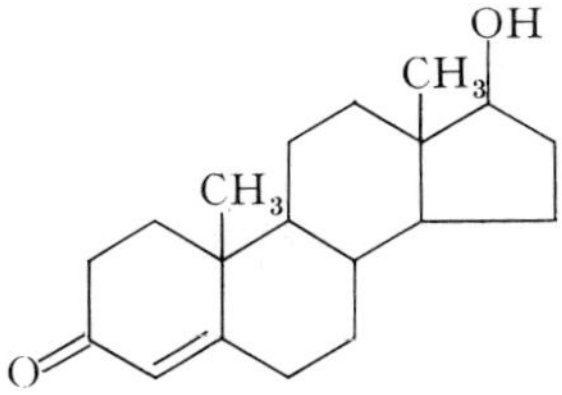

testosterone

The androgens are required for growth and development of the male secondary sexual characteristics, production of sperm, and to promote skeletal and muscular development. They also have an anabolic effect — that is, they stimulate the deposition of tissue and yield a positive nitrogen balance.

Most of the androgens are metabolized to compounds that have a keto group at position 17 on the D ring and are measured in the urine as 17-ketosteroids.

324

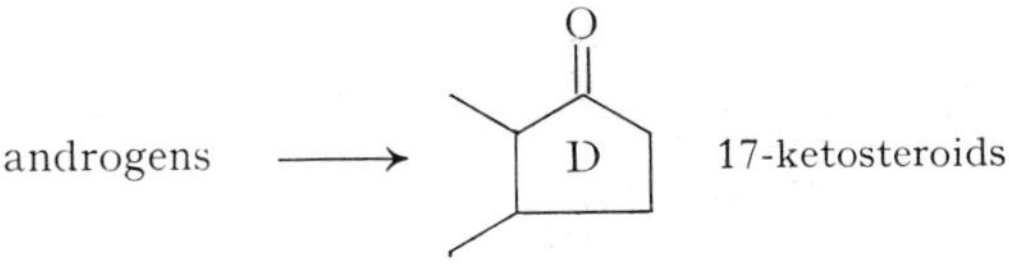

androgens $\longrightarrow$ D 17-ketosteroids

In the male, the adrenal cortex secretes two-thirds of the androgens produced and the testis secretes the remaining one-third. In contrast, the female synthesizes only two-thirds as much androgens as the male, and they are exclusively secreted by the adrenal cortex.

PARATHYROID GLANDS

The parathyroid glands are situated in the neck, two on each side posterior to the thyroid. The glands' protein hormone, parathyroid hormone or parathormone, is important in the regulation of calcium metabolism. This includes the absorption of calcium and phosphorus from the intestines, bone formation, and bone dissolution. The hormone allows the excretion of phosphorus by the kidney and the retention of calcium. It should be recalled that there is a reciprocal relationship between serum calcium and phosphorus concentrations. Therefore, the parathyroids are important not only in controlling calcium per se, but also have a great influence on phosphorus metabolism.

SEROTONIN

The simple organic compound serotonin, which is derived from tryptophan, has been classified by many as a hormone.

325

<chemical structure: tryptophan — indole-CH$_2$-CH(NH$_2$)-COOH> → → <chemical structure: serotonin — 5-HO-indole-CH$_2$-CH$_2$NH$_2$>

tryptophan   serotonin

<chemical structure: 5-HO-indole-CH$_2$COOH>

5-hydroxyindoleacetic acid
(5-HIAA)

Serotonin is involved in blood clotting and normal brain function, but its exact role is not understood. Excessive quantities of serotonin are produced by argentaffin cell tumors (carcinoids). The vasoconstrictive action of serotonin is characteristic of these tumors producing a flushing phenomenon of the skin. Most carcinoids occur along the gastrointestinal tract, most often are benign, and usually can be corrected by surgery. Since blood levels of serotonin are difficult to measure, the degradation product, 5-hydroxyindoleacetic acid (5-HIAA), which is excreted in the urine, is usually quantitated.

This brief discussion of the various hormones is far from complete for there are several more known hormones. This is just a résumé of some of the hormones which are commonly encountered in clinical laboratories.

One of the great difficulties in measuring most of the hormones is that they occur in very small amounts. Very sensitive analytic techniques for hormone assays, including gas chromatography, radioactive isotopes and competitive protein binding, are becoming quite common in many clinical laboratories.

CHAPTER 17

VITAMINS

Vitamins are similar to hormones in that they have profound effects in very small amounts, but differ because all known vitamins are generally simpler compounds and are supplied from external sources. By definition, vitamins are essential, naturally occurring, relatively simple organic compounds required for normal maintenance and development.

Vitamin deficiency signs and symptoms are, in most instances, difficult to evaluate in humans for the individual vitamins. In the majority of clinical cases of vitamin deficiencies there is a deficiency of several of the vitamins. Much of the information available on the subject is obtained from human volunteers on special diets.

For convenience the vitamins may be divided into two broad categories: fat-soluble and water-soluble vitamins.

FAT-SOLUBLE VITAMINS

VITAMIN A

326

$$\text{vitamin A}_1: \text{(cyclohexenyl ring with } CH_3, CH_3, CH_3\text{)}-CH=CH-C(CH_3)=CH-CH=CH-C(CH_3)=CH-CH_2OH$$

The compound is available to the organism in the form of vitamin A or as one of the numerous precursors or provitamins called carotenes. In the intestinal wall, carotenes (compounds containing one or more molecules of vitamin A) are converted to the vitamin. Sources of this vitamin are fish liver oils and yellow and green vegetables.

The greatest effect of a deficiency of vitamin A is impaired vision. Clinically the eyes become dry and ulcerated because of infections. A symptom is night blindness or inability to see in dim light or adapt readily to varying light intensities. Vitamin A serves as a light-sensitive molecule, but the manner in which it converts light energy into nerve stimuli is poorly understood.

VITAMIN D

The active form of vitamin D, 25-hydroxycholecalciferol, is readily synthesized from cholesterol in mammalian organisms by the pathway shown in **327**. This conversion of cholesterol to vitamin D takes place in the skin and is dependent upon radiant energy, particularly from the sun. For this reason, vitamin D is frequently spoken of as the "sunshine vitamin."

Rickets in children and osteomalacia in adults develops from a lack of this vitamin. Defective ossification in rickets causes the bones to become soft and pliable, with deformities such as bow legs, knock-knees, enlargement of the bone ends, and rows of bead-like swellings at the rib junctions. Numerous fractures can result from the brittleness of the bone structure; there may also be deformities and delays in tooth formation.

Osteomalacia presents a different picture because the lesion begins after bone growth is completed. The bones lose a great deal of their calcium and phosphorus and become soft leading to several different types of deformities.

The action of vitamin D is severalfold. It increases the absorption of calcium and phosphorus from the intestinal tract and it also stimulates the deposition of these materials as bone.

VITAMIN E

The structure of vitamin E is somewhat similar to vitamin A. It has been found to play an important role in reproduction in laboratory animals. However, at the present time there is no known requirement for vitamin E by humans.

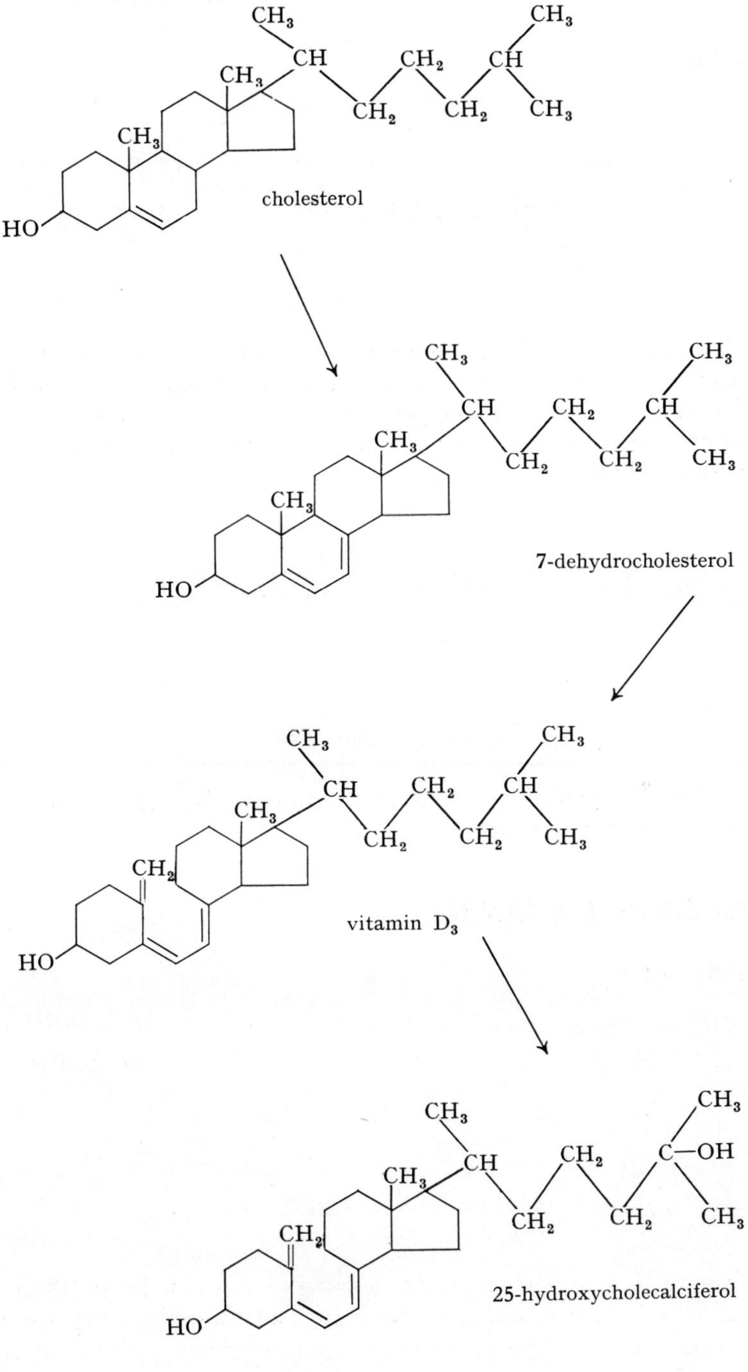

cholesterol

7-dehydrocholesterol

vitamin D₃

25-hydroxycholecalciferol

VITAMIN K

328

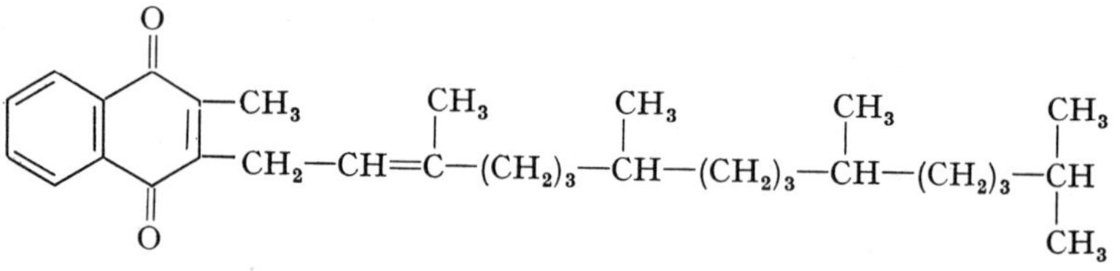

$$\text{vitamin K}_1$$

Structure: 1,4-naphthoquinone with 2-methyl and 3-phytyl side chain:
—CH₃ and —CH₂—CH=C(CH₃)—(CH₂)₃—CH(CH₃)—(CH₂)₃—CH(CH₃)—(CH₂)₃—CH(CH₃)—CH₃

A deficiency of vitamin K results in a marked tendency to bleed profusely from even minor wounds. The liver requires vitamin K to synthesize prothrombin in adequate quantities to prevent hemorrhage. The synthetic analogue of vitamin K, menadione, can replace vitamin K in some instances in therapy.

329

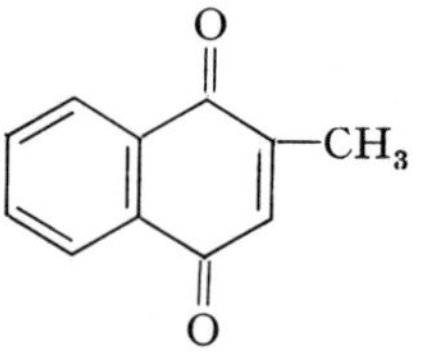

menadione
(2-methyl-1,4-naphthoquinone)

Many different plants are sources of vitamin K, and mammalian liver is an excellent source.

WATER-SOLUBLE VITAMINS

ASCORBIC ACID

330

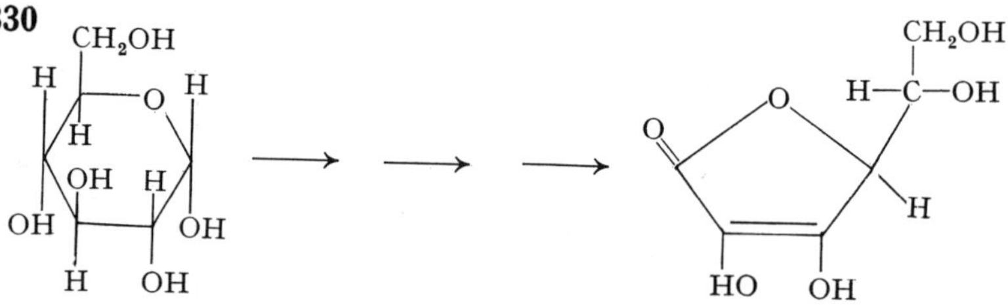

glucose → → → ascorbic acid

Vitamin C is a carbohydrate derivative and can be synthesized from glucose by many animals, but man and the guinea pig are notable exceptions. Vitamin C deficits can cause scurvy. When ascorbic acid is lacking, the intracellular cementing substance becomes deficient and there is hemorrhaging from the mucous membranes in the mouth, from the gastrointestinal tract, as well as from skin, muscle, and other tissue. Also, there is anemia and pain in the joints; the gums are particularly affected, showing ulceration and edema. The primary sources of vitamin C are fruits and some vegetables.

THIAMINE

331

thiamine

Thiamine functions as a coenzyme in the metabolism of α-keto acids. It is phosphorylated to thiamine pyrophosphate, the active form of the vitamin, and functions as a coenzyme for decarboxylase enzymes. These enzymes are involved in decarboxylation of α-keto acids to form the corresponding aldehydes. Deficiencies of thiamine lead to arrested growth, polyneuritis, and a disease in man called beriberi. Thiamine is widely distributed in the plant and animal kingdom; thus both can be considered good sources.

RIBOFLAVIN

332

riboflavin

There is no disease readily recognized as being associated with the deficiency of riboflavin. However, in experimental animals fed diets deficient in riboflavin many dermatologic lesions are found. Riboflavin functions also as a coenzyme. It occurs in flavo proteins, is a component of FMN and FAD, and is concerned with oxidation-reduction reactions. It will be noted that the side chain of riboflavin is ribotol, the alcohol corresponding to ribose. Riboflavin is widely distributed in nature, and plants and animals are both sources for this vitamin.

NIACIN

333

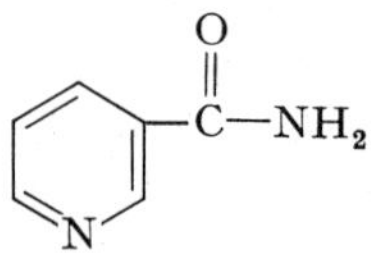

niacin
(nicotinamide)

Niacin is probably one of the simplest vitamins. A condition known as pellagra results from insufficient supplies of niacin, and it was very common among people with low incomes. There were serious problems with this disease in the southern United States until the 1930s. Pellagra is characterized by a number of dermatologic changes as well as keratinization of mucous membranes. Niacin functions as a component of NAD^+ and $NADP^+$, two extremely important coenzymes in oxidation-reduction reactions. The occurrence of niacin is widespread in nature. It is also obtained readily by synthetic techniques.

PYRIDOXINE

334

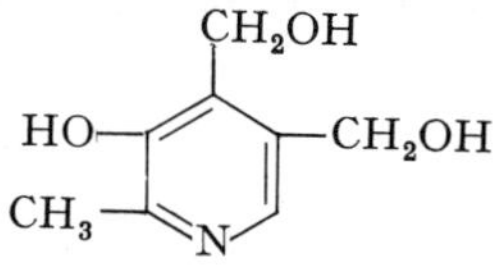

pyridoxine

Pyridoxine, pyridoxal and pyridoxamine, derivatives of pyridine, constitute a family of water-soluble vitamins. Deficiencies in humans are not associated with any known disease; however, laboratory animals with a diet low in this vitamin have dematologic and central nervous system changes. Pyridoxal phosphate functions as a coenzyme for transamination reactions such as SGOT. It is widely distributed in nature so both animals and plants provide excellent sources.

PANTOTHENIC ACID

There seems to be no evidence that panthothenic acid deficits produce disease in humans. Most foodstuffs contain the vitamin and consequently a deficiency is unlikely. Laboratory animals on a diet low in pantothenic acid will have dermatitis, and black rats will have graying of hair. Pantothenic acid is important because it is an intricate component of coenzyme A, which is important in the metabolism of fatty acids and certain other metabolites. Acetyl CoA is the key compound for introducing acetate into the tricarboxylic acid cycle, for the synthesis of fats, for acetylation, and for synthesis of steroids.

FOLIC ACID

335

$$\text{NH}_2\text{—}\underset{\text{OH}}{\underset{|}{\text{pteridine ring}}}\text{—CH}_2\text{—NH—}\underset{}{\text{C}_6\text{H}_4}\text{—}\overset{\text{O}}{\underset{||}{\text{C}}}\text{—NH—CH}\underset{|}{\overset{|}{\text{COOH}}}\text{—CH}_2\text{—CH}_2\text{—COOH}$$

folic acid
(pteroylglutamic acid)

Like pantothenic acid, folic acid serves as a coenzyme. It is, however, involved with the metabolism of one-carbon fragments much like coenzyme A is involved with two-carbon fragments. One-carbon reactions are especially important in amino acid and purine metabolism. In man, folic acid deficiency leads to macrocytic anemia. It is also involved in maintaining normal gastrointestinal absorption.

VITAMIN B_{12}

Vitamin B_{12} has a complex structure and is unique because it contains cobalt and cyanide. Vitamin B_{12} deficiency produces the condition called pernicious anemia. This condition is not necessarily due to a dietary deficiency; it can be caused by the absence of a substance (intrinsic factor) in the stomach which permits absorption of the vitamin. The intrinsic factor is normally secreted by the gastric mucosa; however, in the absence of HCl this factor is not secreted. Thus, analysis of gastric acidity is associated with the diagnosis of pernicious anemia. The action of B_{12} in erythrocyte maturation is not understood, but it is an extremely potent agent. If as little as 1 µg per day is absorbed, satisfactory levels of erythropoiesis are maintained.

BIOTIN

336

$$\text{biotin structure: } \begin{array}{c} O \\ \| \\ C \end{array}$$

```
          O
          ‖
         C
        / \
    H—N   N—H
      |   |
    H—C———C—H
      |   |
    H₂C   C—CH₂—(CH₂)₃—COOH
       \ /
        S
```

biotin

Biotin has long been known to be a growth requirement for certain microorganisms; however it can be demonstrated as a necessary factor for humans only through the use of special diets. Apparently intestinal organisms produce sufficient quantities to meet human requirements. Biotin has been shown to have a role in decarboxylation reactions and to be involved with deamination of certain amino acids.

LIPOIC ACID

337

$$\underset{S\text{————}S}{CH_2\text{—}CH_2\text{—}CH\text{—}(CH_2)_4\text{—}COOH}$$

α-lipoic acid

Lipoic acid is a necessary cofactor for the oxidative decarboxylation of pyruvic and α-ketoglutaric acids. No dietary requirement has been demonstrated for this substance, and it is therefore not considered to be a vitamin by most writers.

VITAMIN SUPPLEMENTS

There are other substances considered to be vitamins or similar to vitamins; however, in most cases little is known about their occurrence and function so they will not be considered here.

Most textbooks and other treatises dealing with vitamins list minimum daily requirements, but to obtain these values is extremely difficult. For obvious reasons, the occurrence of vitamins in food varies tremendously. Thus, it would be impossible to evaluate vitamin requirements unless accurate records of food intake and vitamin content were maintained. Another factor to consider is that cooking and other food processing may destroy some or all of the vitamins present. In the case of vitamin K, man probably has no dietary requirement because intestinal flora usually synthesize adequate amounts. Oral antibiotics may well alter the normal flora to such an extent that a dietary source may be required.

Some pharmaceutical houses and other distributors have capitalized on the uncertainty regarding vitamin requirements by providing the public with numerous vitamin-containing preparations. In a few cases these preparations may be of value, but in most cases they are worthless. Not that they do not contain the amount or types listed on the label, but most ordinary American diets are not vitamin deficient. Generally, an adequate supply of vitamins is assured because a wide variety of foods from good sources is eaten. Only individuals with food idiosyncrasies or those on fasting or special diets need be concerned.

Indeed, the fervor to supply the public with vitamins has extended even to the food processors and distributors. It is quite uncommon today to buy food which is not fortified with some vitamin or vitamins. At first this might be considered a beneficial effort on the part of food suppliers, but vitamins in large amounts are in no way helpful but indeed can be harmful. For instance, even though most vitamins show little toxicity when consumed in large quantities, vitamin D can be toxic to young infants. In Britain, milk suppliers fortified milk with extremely large amounts of vitamin D, which produced toxic effects in children. Now, the addition of vitamins to foods in that country is regulated by the government. As a general rule it can be said that vitamin supplements in reasonable quantities are not harmful but seldom are they helpful, unless the diet is untypical.

CHAPTER 18

PHOTOMETRY

The term photometry as used here will encompass all commonly used clinical laboratory techniques that employ radiant energy measurements: colorimetry, spectrophotometry, flame emission, atomic absorbance, fluorescence, turbidimetry, and nephelometry.

RADIANT ENERGY SPECTRUM

Radiant energy covers a wide frequency spectrum, from the very high frequencies (gamma, cosmic, and x-rays) to the very low frequencies, such as radio waves. The radiant energy spectrum is as follows:

TABLE 18-1. RADIANT ENERGY SPECTRUM

Radiation	Wavelength (nm)
γ-rays	0.1
x-rays	0.1 to 14.0
ultraviolet	14 to 400
visible	400 to 800
infrared	800 to 400,000
radio waves	4×10^5 to 3×10^{13}

There is an inverse relationship between wavelength and frequency.

338
$$F = \frac{c}{\lambda}$$

where F = frequency,
 c = velocity, 3×10^{10} cm/second
 λ = wavelength

The amount of energy in the radiation is directly related to frequency.

339 $$E = hF$$

where E = energy
h = Plank's constant, 6.6×10^{-27} erg second

X-rays are in the high-frequency area of the radiant energy spectrum, and as is well known, they have sufficient energy to penetrate tissue and form x-ray images. This is not true for visible light with lower frequencies, which has less energy and does not penetrate tissue.

Currently in clinical laboratories only a small segment of the spectrum, the ultraviolet and visible regions, is generally used for quantitation.

ABSORBANCE PHOTOMETRY

Quantitation is accomplished by comparing the amount of radiant energy (light for convenience) absorbed by a standard sample of known concentration with that of a sample with an unknown concentration.

An analogy is a coffee drinker's ability to recognize a cup of strong coffee by its appearance (unknown). The darker the color, the greater the light absorbance. For a standard in this instance, the coffee drinker must rely on his memory.

An early type of absorbance photometer is shown schematically in Figure 18-1. Light from a source, which can be an ordinary elec-

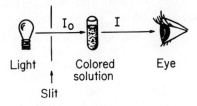

Fig. 18-1. Early photometer.

tric light, is directed through a slit or aperture to a tube of colored solution. This is called incident light (I_0). Some of the light is reflected and refracted and some is absorbed by the sample and the tube. A portion of the light will pass through the solution and reach the eye. This is called transmitted light (I).

As the experienced student might well suspect, there are some mathematical formulas involved with quantitation. These formulas are derived from the Lambert and Beer laws.

The Lambert law states that for a given quantity of incident light the amount transmitted through a solution is inversely related to the depth of the colored solution. In more simple terms, more light will be transmitted by a small tube containing a colored solution than through a much larger one (Fig. 18-2).

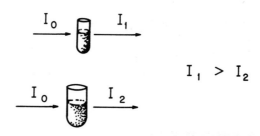

Fig. 18-2. Demonstration of Lambert law.

The Lambert law relates these factors as

340 $\qquad$ (a) $\qquad \dfrac{I}{I_0} \propto \dfrac{1}{d}$

$\qquad$ (b) $\qquad \dfrac{I}{I_0} = a^{-d}$

where d = depth of solution,
a = constant for particular colored solution.

The Beer law relates incident and transmitted light to the concentration of colored materials

341 $\qquad$ (a) $\qquad \dfrac{I}{I_0} \propto \dfrac{1}{c}$

$\qquad$ (b) $\qquad \dfrac{I}{I_0} = a^{-c}$

where c = concentration

Combining both mathematical expressions **340** (b) and **341** (b) yields the Beer-Lambert law (**342**).

342 $\qquad \dfrac{I}{I_0} = a^{-dc}$

The fraction I/I_0 is defined as transmittance (T),

343 $$T = \frac{I}{I_0} = a^{-dc}$$

Percent transmittance (% T) therefore is,

344 $$\% T = \frac{I}{I_0} \times 100 = T \times 100$$

Consider an experiment where the transmittances of colored solutions of the same material, but with different concentrations, were measured using the same tube and under the same conditions. The transmittance measurements were made in the same tube, so the depth d of the solution is constant. The other variable, a, is also constant because the same colored material is measured under the same conditions each time. The transmittance measurements plotted against concentration produce the curve shown in Figure 18-3.

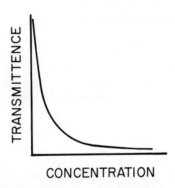

Fig. 18-3. Relationship of transmittance and concentration.

It is obvious that this is a difficult function with which to work. If such a plot is used to find the concentration of unknowns, many points would be required to describe such a function. For this reason, transmittance measurements are avoided whenever possible.

By a little manipulation of **342** it is possible to derive a very simple and useful expression for photometry. Writing **342** as a log function yields **345**.

345 $$\log \frac{I}{I_0} = - dc \log a$$

When the measurements are made at a constant depth and for the same colored materials, d and a are constant. If a is a constant, then log a is also constant and the product of two constants is just another constant. Therefore

346
$$d \log a = K$$
$$\log \frac{I}{I_0} = -cK$$

Multiplying both sides of the equation by -1 gives

347
$$-\log \frac{I}{I_0} = cK$$

The negative sign is removed from the left side of the equation by inverting that portion. This mathematical maneuver is possible because numbers are divided by subtraction of their logarithms. For example

348
$$-\log \frac{I}{I_0} = -(\log I - \log I_0) = -\log I + \log I_0 = \log \frac{I_0}{I}$$

The equation can now be written as

349
$$\log \frac{I_0}{I} = cK$$

Absorbance (A) is defined as

350
(a) $\quad A = \log \frac{I_0}{I}$

(b) $\quad A = cK$

Measurements of A are directly proportional to c, and a plot of A versus c yields a useful straight line with an intercept of zero. The slope of the line is, of course K, (Fig. 18-4).

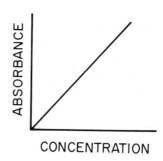

Fig. 18-4. Relationship of absorbance and concentration.

ABSORBANCE AND TRANSMITTANCE LIMITS

Using the following formulas for reference, consider the limits as the value of I changes with respect to I_0.

350 (a) $$A = \log \frac{I_0}{I}$$

343 $$T = \frac{I}{I_0}$$

344 $$\% T = \frac{I}{I_0} \times 100$$

As I approaches I_0 ($I \rightarrow I_0$), then $A \rightarrow 0$, $T \rightarrow 1$, and $\% T \rightarrow 100\%$; and when $I = I_0$, $A = 0$, $T = 1$, and $\% T = 100$. As $I \rightarrow 0$, $A \rightarrow \infty$, $T \rightarrow 0$, and $\% T \rightarrow 0$. In practice, instruments for measuring absorbance are limited to 2 or 3, so that for practical purposes $A_{\text{lim}} = 2$ or 3.

INTERCONVERSION OF ABSORBANCE AND TRANSMITTANCE

It is necessary at times to collect transmittance data, but since it is usually difficult to work with, it is generally converted to absorbance. The following equations are used for the conversion.

351 $$A = -\log T = \log \frac{1}{T}$$

352 $$A = -\log \frac{\% T}{100} = -(\log \% T - \log 100) = -\log \% T + \log 100$$

353 since $\log 100 = 2$

then

$$A = 2 - \log \% T$$

The conversion of A to T is simply

354 $$T = \frac{1}{\text{antilog } A}$$

ABSORBANCE PHOTOMETERS

All photometers consist of four essential elements: radiant energy source, sample, photodetector, and readout device (Fig. 18-5).

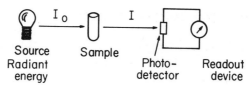

Fig. 18-5. Simple photometer.

RADIANT ENERGY SOURCES

Different types of radiant energy sources are required to provide different portions of the spectrum. Lamps with tungsten filaments are commonly used for absorbance measurements in the visible region of the spectrum, because tungsten filaments emit radiation covering the entire visible range, 400 to 800 nm. Gas discharge lamps using hydrogen or deuterium provide emission from 200 to 400 nm for ultraviolet (UV) absorbance. Other gas-filled tubes can be used for special applications. Mercury lamps provide several wavelengths of intense emission (bright line spectra) that are very useful in calibrating photometers in the visible and ultraviolet regions. The discharge lamps have the disadvantage of requiring a separate power supply.

Most photometric instruments have an optical system located between the lamp and sample compartment. This is not shown in Figure 18-5 but may consist of a single, simple collimating lens to a rather complicated lens and mirror system. Slits or apertures are used to limit the amount of light entering the sample compartment.

Photometers also have devices to select the portion of the spectrum that enters the sample compartment. Most radiant energy sources emit a broad spectrum, but usually only a narrow segment can be used effectively. There are two principle reasons for a narrow spectral band. One, most substances have greater absorbance at specific wavelengths. For example, if absorbance measurements are recorded for a substance at each wavelength (shown in Figure 18-6 as a solid line), it is seen that the material has maximum absorbance at 540 nm.

Radiation of this substance with light at 540 nm (shown in Figure 18-6 as a heavy vertical line) would result in the absorbance of a large portion of this light. If this same solution is irradiated with light having wavelengths from 400 to 800 nm (shown as a dotted line), the material would absorb only a small amount of the total radiant energy. The use of a narrow band of light, then, allows greater absorbance and thus greater differences for better quantitation.

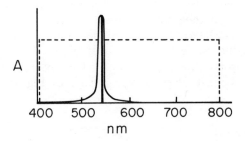

Fig. 18-6. Relationship of absorbance and wavelength for a substance with maximum absorbance near 540 nm.

Another reason for using a narrow portion of the spectrum is that it allows selective absorbance measurements in instances where there is more than one absorbing substance. The absorbance spectra of a solution containing two substances, A and B, are shown in Figure 18-7.

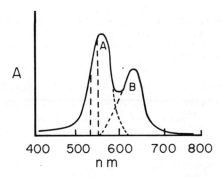

Fig. 18-7. Absorbance-wavelength curve for a solution containing two absorbing substances (A and B).

If the absorbance of the solution were measured at wavelengths from 400 to 800 nm, the total absorbance for A and B would be obtained. However, if a narrow band of 520 to 550 nm was used (dashed lines in Figure 18-7), only the absorbance of A would be measured. The use of narrow parts of the spectrum allows selective absorbance measurements and the measurements of selected absorbing materials.

Filters, Prisms and Gratings

There are four commonly used methods for isolating portions of the spectrum. The simplest are selective transmission filters constructed of colored glass or other colored transparent materials. A dilute solution of hemoglobin appears red because all the visible light except red is absorbed by the hemoglobin. Transmission filters operate in the same way in that one constructed of red glass transmits red light and absorbs the other colors. The absorbance spectrum of a typical colored glass filter is represented in Figure 18-8.

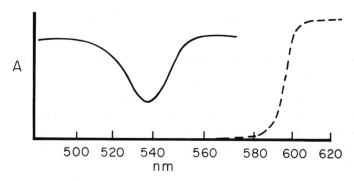

Fig. 18-8. Absorbance-wavelength characteristics for a colored glass filter showing maximum transmittance near 540 nm (*solid line, left*) and the characteristics of a cutoff filter that does not transmit at wavelengths greater than 600 nm (*dashed line, right*).

Another type of transmission filter is the cutoff. These filters allow a portion of the spectrum to pass but absorb the rest. An example is seen in Figure 18-8 shown as a dashed line. Here the filter transmits well at about 580 nm, and at 600 nm transmits almost no light. Cutoff filters are very useful to completely exclude a portion of the spectrum.

Neutral density filters are transmission filters that have a nearly uniform absorbance over a wide range of wavelengths. They are interposed between the radiant energy and sample to reduce the light intensity, or, in some instances, they can be used in place of the sample as an absorbance standard.

A better type of filter is an interference filter. The operating principle of these filters is a complex physical phenomenon that allows them to transmit very narrow bands. A disadvantage of these filters is that the amount of light they transmit is low compared with glass filters.

Prisms are also used to isolate spectra. When a prism is irradiated with white light, the emerging light will be dispersed into an infinite number of colors, with the light of the shortest wavelength (greatest frequency) refracted the most. The dispersion of white incident light is schematically represented in Figure 18-9 but only

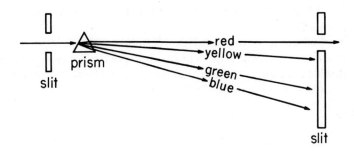

Fig. 18-9. Dispersion of white incident light by a prism.

the four major colors are shown for clarity. For the isolation of red light, a slit would be placed as shown to allow passage of the red light but blocking the other wavelengths. In instruments employing prisms, wavelength selection is accomplished by rotating the prism so that the desired wavelengths pass through the slit.

Diffraction gratings provide another method of dispersing light. A grating consists of a flat or curved plane traversed with a large number of very small parallel grooves as illustrated in Figure 18-10. Wavelength selection is accomplished by rotating the grating so that the desired wavelength passes through the exit slit.

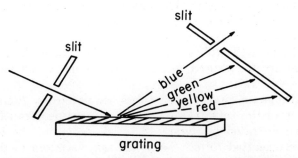

Fig. 18-10. Dispersion of white incident light by a diffraction grating.

One of the number of factors that influences choice of a wavelength selector is the type of radiant energy to be measured. Since ordinary glass is nearly opaque to ultraviolet radiation, glass transmission, cutoff and neutral density filters, and glass lenses cannot be used in this region of the spectrum. This is also true for glass prisms. Quartz is most frequently used for lenses and prisms when ultraviolet radiation is used for absorbance measurements. Gratings do not present a problem in this regard because they do not transmit light.

Another component found in photometers that have the source of radiant energy near the sample compartment is a heat shield. Heat shields are made from transparent materials that absorb infrared.

CUVETS

The containers that hold the sample in photometry are called cuvets. They can be obtained in a variety of sizes, shapes, and materials, such as quartz, glass, and plastic. Many instruments can accommodate ordinary test and culture tubes, which are quite adequate for many applications. When round cuvets are used a couple of factors must be considered, one of which is the geometry. If different tubes are used they must provide the same depth for the

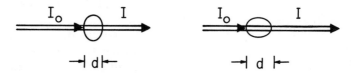

Fig. 18-11. Effect of position of nonround cuvets on depth of light path.

samples. Tubes can be quickly selected, "matched," for consistent depth by placing the same colored solution in a large group of similar tubes. Those that demonstrate a uniform absorbance (mean $\pm$ 2% of mean) can be grouped as a matched set. If the instrument has a variable slit then the tubes must be matched for each slit width at which they are to be used, because the geometry can change with changes in slit width.

Another factor is that most tubes are not really round, so that the position of the tube in the sample compartment can be critical. This is demonstrated in Figure 18-11 as seen from the top of the tubes in exaggerated form. To avoid this variation in depth with tube rotation, cuvets are marked (etched) so that they can always be positioned the same way. Any mark on the tube that is permanent, such as a trade name or etching, can serve as a position reference. A large percentage of the commercially available culture tubes can be selected to form one or two matched sets.

Round cuvets are difficult to use except at fixed wavelengths. Many instruments have variable slit widths to accommodate the energy change with wavelength changes. As indicated in **339,** energy is related to frequency, and as the frequency of the radiation decreases, so does its energy. Further, frequency and wavelength are inversely related; thus, as the wavelength becomes longer the energy becomes less. For example, at longer wavelengths — the red portion of the visible spectrum — it is sometimes necessary to introduce more light into the sample compartment to provide sufficient energy for good performance of the photodetector. This is not always the case, however, because some photodetectors are more responsive to red light than they are to the shorter wavelengths. Regardless, some instruments do have variable slit widths, and when the width changes so does the geometry of the cuvet. For the same cuvet, there is a difference in the area traversed by a light beam from a narrow to a wide slit (Fig. 18-12). The narrow light

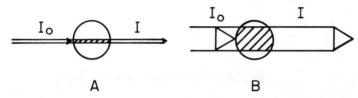

Fig. 18-12. Change in geometry due to change in slit width for round cuvet.

beam in Figure 18-12 A essentially traverses an area that is a long thin rectangle. In B, with a wide slit, the area no longer approximates a rectangle.

The problems inherent in round cuvets can be avoided with rectangular ones. The depth remains the same regardless of slit width, as illustrated in Figure 18-13.

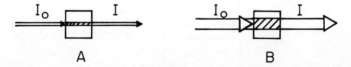

Fig. 18-13. Constant depth of light path with change in slit width for rectangular cuvet.

Regardless of the size or shape of cuvets they must always be meticulously clean. Any substance on the surface of the cuvet, including fingerprints, can absorb, refract, or reflect light, which would lead to erroneous absorbance.

PHOTODETECTORS

The only practical way to measure radiant energy is to convert it to electrical signals. Devices which convert one form of energy to another are called transducers. In photometry they are referred to as photocells or photodetectors. There are two basic types: photovoltaic and photoresistive. The photovoltaic is a solid state device that generates a difference in potential proportional to the illumination. A photovoltaic circuit is schematically shown in Figure 18-14.

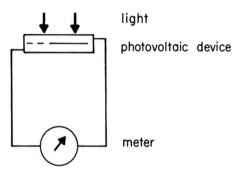

Fig. 18-14. Photovoltaic circuit.

The output of photovoltaic detectors is sufficient so that no external power source is required if a sensitive galvanometer is used. These units are rugged, inexpensive, and have a nearly linear response to light. One disadvantage, however, is instability ("fatigue") after long exposure to light. Another disadvantage is that the use is restricted to the visible portion of the spectrum.

The second detection device is photoresistive or photoconductive. It allows current flow in proportion to illumination (Fig. 18-15).

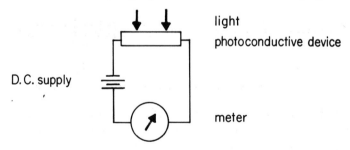

Fig. 18-15. Photoconductive circuit.

The photoconductive device, in many instances, is incorporated into an evacuated glass or quartz envelope. In this case, they are called phototubes or vacuum phototubes. Some are constructed to amplify (photomultipliers) the electrical signals up to one million times, which allows detection of very low light intensities. Phototubes can be used to detect in the ultraviolet as well as in the visible range. A disadvantage of photoconductive devices is that an external power supply is required.

READOUT DEVICES

In Figs. 18-14 and 18-15 the current flow was indicated by a meter. Readout devices can vary from very simple galvanometers to analogue to digital converters interfaced with digital computers. Generally, the photometers used in clinical laboratories employ a meter, digital display, or a strip chart recorder.

If the readout instrument is a galvanometer with a linear scale, the values obtained are transmittance values. With a logarithmic scale, the values are absorbances or an equally useful function of absorbance. Several commercial instruments still provide % T readouts in spite of the fact that in the majority of instances absorbance values are desired. Some instruments can be adjusted so that the readout is in concentration units rather than in terms of absorbance.

The radiant energy source, sample, photodetector, and readout devices as shown in Figure 18-5 are found in all photometers. A more realistic diagram including the items discussed here would appear as shown in Figure 18-16.

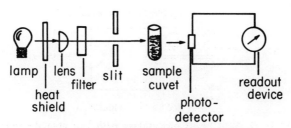

Fig. 18-16. Diagram of simple photometer.

ZERO ABSORBANCE AND TYPES OF PHOTOMETERS

It is necessary to establish a zero absorbance response for the instrument before absorbance measurements are made. Some light is reflected, refracted, and absorbed by any cuvet and the solution it contains. To define zero absorbance for a particular situation, the cuvet is filled with what is commonly called a "blank solution."

There are actually several different types of blanks and several ways that blanks can be used. The two blanks most commonly used are reagent and sample. A reagent blank is used to correct for the absorbance of reagents, and a sample blank is used to correct for extraneous sample absorbance.

When the cuvet containing the blank solution is placed in the sample compartment, the instrument is adjusted so that the readout device indicates zero absorbance. This is usually accomplished with a variable resistor in the measuring circuit of single beam photometers or by regulating the amount of light reaching the reference photocell in some split beam photometers. Some instruments also require adjustment of infinite absorbance, $0 \% T$. Various samples can now be introduced and the corresponding absorbance values obtained.

Single beam photometers have several advantages (Fig. 18-16): (1) Measurements can be made quickly once the zero absorbance adjustment is made. (2) It employs only one photodetector, which eliminates the need for selected compatible detectors required for instruments with more than one. A major disadvantage is that any condition change in the instrument, such as light intensity, photodetector response, or readout devices, could go unnoticed. Therefore, it requires continuous monitoring with blank and standard solutions.

Split beam or "dual beam" instruments provide continuous monitoring of a reference (Fig. 18-17). For simplicity, the compo-

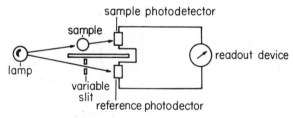

Fig. 18-17. Split beam photometer with two photodetectors.

nents between the lamp and sample compartment have been omitted. To adjust to zero absorbance a blank solution is placed in the sample compartment, and the variable slit, regulating the amount of light falling on the reference photodetector, is adjusted so that the readout device indicates zero absorbance. The instrument is now ready for measurements of sample absorbance. An advantage with these photometers is that both the reference and sample detectors are affected by any variation in source (lamp) intensity or measuring system. For good performance the photodetectors must have similar responses at all wavelengths and all light intensities.

The same approach is applicable in more sophisticated instruments with a single photodetector, as seen in Figure 18-18. Beam

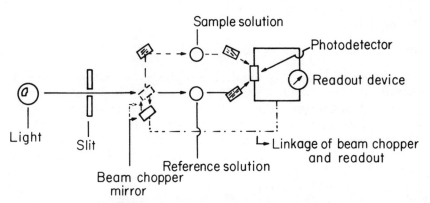

Fig. 18-18. Split beam photometer whith a single photodetector.

choppers are used to divert the light alternately through the sample and reference solution. The mechanical chopper is synchronized with the readout device so that the difference in absorbance between the two light paths is measured. It has the advantage of a constant reference absorbance solution (blank) as well as only a single photodetector.

ABSORBANCE PHOTOMETRY AND QUANTITATION

It was shown in **350** (b) that $A = ck$. In the previous discussion it was indicated that A was a relative, rather than an absolute quantity. The blank solution is used to define zero absorbance for the particular assay condition. Quantitation is accomplished by comparing the absorbance of a standard solution with that of an unknown solution under identical conditions. The absorbance of the standard solution would be related to the concentration, c_s, by

355
$$A_s = c_s K$$

Similarly, the absorbance (A_u) and concentration (c_u) of the solution with the unknown concentration would be

356
$$A_u = c_u K$$

Dividing **356** by **355**

357   (a) $\quad \dfrac{A_u}{A_s} = \dfrac{c_u K}{c_s K}$

or

      (b) $\quad \dfrac{A_u}{A_s} = \dfrac{c_u}{c_s}$

Solving for c_u

358 $\quad\quad\quad\quad\quad\quad c_u = \dfrac{A_u}{A_s} \times c_s$

This simple expression is applicable if the standard and unknown solution are treated identically. For example, if 100 µl of serum and 100 µl of glucose solution, 100 mg/dl, were treated identically with reagents that produce a color with glucose, and the glucose standard solution had an absorbance of 0.197 and the serum an absorbance 0.154, the glucose concentration of the serum would then be

359 $\quad\quad\quad\quad\quad c_u = \dfrac{0.154}{0.197} \times 100 \text{ mg/dl} = 78.7 \text{ mg/dl}$

If the standard and unknown solutions are manipulated differently, or if it is desirable to change concentration units, then a conversion factor is required.

360 $\quad\quad\quad\quad\quad\quad c_u = \dfrac{A_u}{A_s} \times c_s \times f$

where f = conversion factor

The use of such a factor can be illustrated by the following example for the estimation of χ. First, 1 ml of urine was placed in a 100 ml volumetric flask, diluted to volume with water, and mixed. Then 2 ml of the diluted urine and 2 ml of a standard solution of χ, containing 10 µg χ/2 ml, were used for the colorimetric assay. The absorbance of the standard solution (A_s) was 0.210 and the absorbance of the unknown solution (A_u) 0.096. The problem is to calculate the total amount of χ in the urine with a total volume of 1,937 ml. From **360**

361 $\quad\quad\quad\quad\quad c_u = \dfrac{0.097}{0.210} \times 10 \text{ µg} \times f$

The value of f is found as follows. The volume of urine actually analyzed was

362 $\quad\quad\quad\quad\quad \dfrac{1 \text{ ml}}{100 \text{ ml}} \times 2 \text{ ml} = 0.02 \text{ ml}$

where 1 ml = amount of urine diluted
       100 ml = volume of diluted urine
       2 ml = amount of diluted urine used in assay

Since the total amount of χ is the value desired, f becomes

363
$$f = \frac{1937}{0.02} = 96{,}850$$

Incorporating **363** into **361** provides

364
$$c_u = \frac{0.097}{0.210} \times 10.0\ \mu g \times 96{,}850$$

$$= 447{,}350\ \mu g / 1{,}937\ ml \longrightarrow 450{,}000\ \mu g\ total$$

If it was desired to convert micrograms to milligrams this factor could also be incorporated into the factor by dividing it by 10^3, the number of micrograms per milligram.

365
$$f' = \frac{f}{10^3\ \mu g/mg} = \frac{96{,}850}{10^3\ \mu g/mg} = 96.85\ mg/\mu g$$

The calculation would then be

366
$$c_u = \frac{0.097}{0.210} \times 10\ \mu g \times 96.85\ mg/\mu g$$

$$= 447\ mg/1{,}937\ ml \longrightarrow 450\ mg\ total$$

Some analysts prefer to fill a cuvet with water to adjust the instrument to zero absorbance and then measure the absorbance of the blanks. If this approach is used the absorbance of the blanks (A_b) must then be subtracted from the absorbance of the unknown and standards.

367
$$c_u = \frac{A_u - A_b}{A_s - A_b} \times c_s \times f$$

CALIBRATION OR STANDARDIZATION CURVES

There are several substances commonly measured in clinical laboratories that are unstable, and as a result, a standard solution cannot be prepared to use with each analysis. A standardization or calibration curve can be used in this situation. To establish such a curve, standard solutions are prepared under optimum conditions and are assayed. It is frequently desirable to dissolve the standard material in the biologic material being assayed, particularly if the biologic sample is serum or plasma and deproteinization is in the assay procedure. Each standard solution is carefully analyzed at least in duplicate, and the absorbance values are plotted against the corresponding values of concentration. Subsequently, when samples are analyzed their concentrations are found by use of the calibration curve. If more than one photometer is to be used for a given assay, each instrument must be calibrated. One calibration curve is not adequate because instrument response is different for each and every instrument.

Whenever possible calibration curves should be avoided, because when using them one assumes that there is no change in reagents, instrument response, or technique from the time the curve was established. Such an assumption is optimistic to say the very least. Instrument response can be monitored by measuring the absorbance of a stable solution absorbing at the same wavelength as used for the assay, each time an analysis is performed. If there is deviation of more than ± 1 to 2% from the time the calibration curve was established, it suggests that recalibration is required. The only way to monitor the other factors is with a good quality control program.

MOLAR AND SPECIFIC ABSORPTIVITY

The molar absorptivity of a substance is defined as the absorbance of a one-molar solution at specified depth and wavelength. In actual practice it is seldom possible to measure the absorbance of a one-molar solution because the absorbance is too great to measure. Absorbance measurements are made on less concentrated solutions, and the corresponding absorbance of a one-molar solution is found by calculation.

If other factors, such as pH, temperature, or solvent, affect the absorbance, then they too must be specified. An example would be the molar absorptivity of bilirubin.

368 $\qquad M \sqcap_{10 \text{ nm}}^{453 \text{ nm}} = 5.93 \times 10^4$ (chloroform)

where $M \sqcap$ = molar absorptivity
 453 nm = wavelength of measurement
 10 nm = depth of solution
(chloroform) = solvent

The specific absorptivity, $s\sqcap$, is defined as the absorbance of a 1 g/dl solution at specified depth and wavelength. For bilirubin this would be

369 $\qquad s \sqcap_{10 \text{ nm}}^{453 \text{ nm}} = 1.01 \times 10^3$

Specific absorptivity is used when the molecular weight of the substance is not known, as in the case for many proteins, nucleic acids, polysaccharides, and other macromolecules.

Absorptivities are additional properties of materials that are often useful for quantitation and identification. However, some care must be exercised in their application in that there is considerable variation in the values of absorptivities when measured with different instruments.

Absorptivities can be used to find the concentration of two substances with overlapping absorbance spectra. Consider Figure 18-19 where the dotted curve shows the absorbance spectrum of substance P at a concentration of $10^{-2} M$; the dashed line is the spectrum for Q at $10^{-3} M$; the solid line is the spectrum for a solution which is $10^{-2} M$ P and $10^{-3} M$ Q. Assume that it was known that a solution contained P and Q, but that the concentrations of each were unknown, and it was desired to determine the concentration of each. The procedure is to determine the spectra of P and Q as shown in Figure 18-19, then calculate the absorptivities of each at

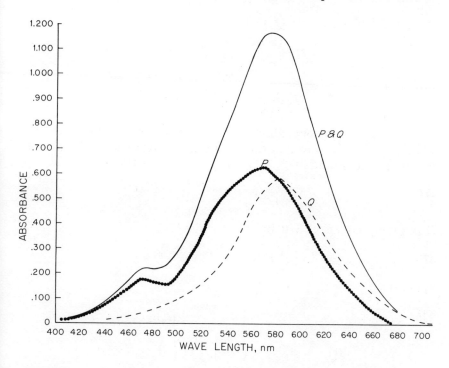

Fig. 18-19. Absorbance spectra of solutions of substance P (*dotted line*) and substance Q (*dashed line*) and a solution containing both P and Q (*solid line*).

two wavelengths. The number of absorptivities required is equal to the number of components in the system. The wavelengths chosen to calculate the absorptivities in this example are 525 and 590 nm. The absorbance of P at 525 nm ($A_p^{525 \text{ nm}}$) is 0.418. All the measurements were made at a depth of 10 mm in a rectangular cuvet, so designation of depth will be disregarded. From the definition of molar absorptivity

370 $$M\boxed{}_p^{525 \text{ nm}} = \frac{A_p^{525 \text{ nm}}}{c_p}$$

where c_p = molar concentration of P.

371 $$M\boxed{}_p^{525 \text{ nm}} = \frac{0.418}{10^{-2}} = 41.8$$

The molar absorptivity of P at 590 nm, where the absorbance is 0.530 is

372 $$M\boxed{}_p^{590 \text{ nm}} = \frac{0.530}{10^{-2}} = 53.0$$

Similar calculations for Q yield

373 (a) $$M\boxed{}_q^{525 \text{ nm}} = \frac{0.180}{10^{-3}} = 180$$

and

(b) $$M\boxed{}_q^{590 \text{ nm}} = \frac{0.555}{10^{-3}} = 555$$

In most situations absorbances at the same wavelength are additive. This is shown as the solid line in Figure 18-19 for a solution of $10^{-2}\,M$ P and $10^{-3}\,M$ Q. It follows then that

374 $$A_{p+q}^{525 \text{ nm}} = A_p^{525 \text{ nm}} + A_q^{525 \text{ nm}}$$

Multiplying equation **370** by c_p provides

375 $$(c_p)\left(M\boxed{}_p^{525 \text{ nm}}\right) = A_p^{525 \text{ nm}}$$

Similarly

376 (a) $$A_q^{525 \text{ nm}} = \left(M\boxed{}_q^{525 \text{ nm}}\right)(c_q)$$

(b) $$A_p^{590 \text{ nm}} = \left(M\boxed{}_p^{590 \text{ nm}}\right)(c_p)$$

(c) $$A_q^{590 \text{ nm}} = \left(M\boxed{}_q^{590 \text{ nm}}\right)(c_q)$$

Substituting **375** and **376** (a) into **374**

377 $$A_{p+q}^{525\,nm} = \left[\left(M_p^{525\,nm}\right)(c_p)\right] + \left[\left(M_q^{525\,nm}\right)(c_q)\right]$$

This provides an equation relating absorbance to concentration, but it contains two unknowns. The absorbances of pure solutions of P and Q at known concentrations were measured and $M_p^{525\,nm}$ and $M_q^{525\,nm}$ were calculated. Since a single equation with two unknowns can present a problem, another approach is desired. Consider absorbance of the solutions of P and Q at 590 nm.

378 $$A_{p+q}^{590\,nm} = A_p^{590\,nm} + A_q^{590\,nm}$$

Using equations **376** (b) and **376** (c) the following will result

379 $$A_{p+q}^{590\,nm} = \left[\left(M_p^{590\,nm}\right)(c_p)\right] + \left[\left(M_q^{590\,nm}\right)(c_q)\right]$$

As before, the equation has two unknowns. Now, however, there are two different equations with the same unknowns and they can be solved simultaneously. Multiplying **377** by

$$-\frac{M_p^{590\,nm}}{M_p^{525\,nm}}$$

380 $$-\left(\frac{M_p^{590\,nm}}{M_p^{525\,nm}}\right)\left(A_{p+q}^{525\,nm}\right) = \left[-\left(\frac{M_p^{590\,nm}}{M_p^{525\,nm}}\right)\right]\left[\left(M_p^{525\,nm}\right)(c_p)\right]$$
$$+ \left[-\left(\frac{M_p^{590\,nm}}{M_p^{525\,nm}}\right)\right]\left[\left(M_q^{525\,nm}\right)(c_q)\right]$$

Canceling yields

381 $$-\left(\frac{M_p^{590\,nm}}{M_p^{525\,nm}}\right)\left(A_{p+q}^{525\,nm}\right) = -\left(M_p^{590\,nm}\right)(c_q)$$
$$-\left(\frac{M_p^{590\,nm}}{M_p^{525\,nm}}\right)\left(M_q^{525\,nm}\right)(c_q)$$

Adding **379** and **381** provides

382
$$\left(A_{p+q}^{590\,nm}\right) - \left(\frac{M|_p^{590\,nm}}{M|_p^{525\,nm}}\right)\left(A_{p+q}^{525\,nm}\right) = \left(M|_q^{590\,nm}\right)(c_q)$$
$$- \left(\frac{M|_p^{590\,nm}}{M|_p^{525\,nm}}\right)\left(M|_q^{525\,nm}\right)(c_q)$$

Solving for c_q

383
$$c_q = \frac{\left(A_{p+q}^{590\,nm}\right) - \left(\dfrac{M|_p^{590\,nm}}{M|_p^{525\,nm}}\right)\left(A_{p+q}^{525\,nm}\right)}{\left(M|_q^{590\,nm}\right) - \left(\dfrac{M|_p^{590\,nm}}{M|_p^{525\,nm}}\right)\left(M|_q^{525\,nm}\right)}$$

Finally, after all that maneuvering, one can arrive at an equation that has a single unknown. Substituting into **383** the measured values listed

384
$$A_{p+q}^{590\,nm} = 1.085$$
$$M|_p^{590\,nm} = 53.0$$
$$M|_p^{525\,nm} = 41.8$$
$$A_{p+q}^{525\,nm} = 0.598$$
$$M|_q^{525\,nm} = 180$$
$$M|_q^{590\,nm} = 555$$

$$c_q = \frac{1.085 - \left(\dfrac{53.0}{41.8} \times 0.598\right)}{555 - \left(\dfrac{53.0}{41.8} \times 180\right)}$$

$$c_q = \frac{1.085 - 0.758}{555 - 228} = \frac{0.327}{327} = 10^{-3}\,M$$

Now that a value has been obtained for c_q the rest is easy. Using

$$c_p = \frac{A_{p+q}^{525\,nm} - \left(M\epsilon_q^{525\,nm}\right)(c_q)}{M\epsilon_p^{525\,nm}}$$

$$= \frac{0.598 - (180 \times 10^{-3})}{41.8}$$

$$= \frac{0.598 - 0.180}{41.8} = \frac{0.418}{41.8} = 10^{-2}\,M$$

This specific example of a generally applicable technique illustrates a very convenient assay procedure for absorbing materials. Six quick measurements and a little algebra were all that was required. The only difficulty with this technique is that the number and identity of all absorbing components must be known to obtain valid results.

COLORIMETERS AND SPECTROPHOTOMETERS

In general, absorbance photometers are divided into two groups: filter photometers and spectrophotometers. Filter photometers are frequently limited to the visible spectrum and are often called colorimeters. A major disadvantage is that the wavelengths are limited by the number of available filters. A sensitive colorimeter using interference filters, however, can approach the capability of most spectrophotometers. This type of instrument has the advantage of being less expensive than spectrophotometers, which often means that more can be available for the convenience of laboratory workers.

Spectrophotometers employ prisms or gratings for spectral isolation thus providing the advantage of a continuous selection of wavelengths. This makes it possible to establish absorbance curves as shown in Figure 18-19. Of course, the same could be done with filter instruments, but it would require a great number of different filters. Another advantage is that the wavelength selection permits a narrow range of wavelengths (narrow band width), thus more selective absorbance measurements are possible. For instance, in Figure 18-7 the absorbance of A could be measured without interference from B because the radiation has a narrow band. It is obvious that the narrower the band the more selective the measurements can be.

A decision as to whether one good spectrophotometer or several filter instruments should be purchased is usually based on money. With this problem answered, the question of which commercially available instrument to obtain must be resolved. Fortunately, there is an increasing number of photometer manufacturers, and the competition tends to promote the development of better instruments. The numerous choices make selection more difficult. Instrument selection should be based on the instruments' performance with a particular assay. In other words, it would be hardly necessary to employ a photometer that has reproducibility of $\pm 0.1\%$ absorbance when the rest of the assay has a variation of $\pm 20\%$. Frequently, rather inexpensive instruments are found to provide very satisfactory results. The only real way to find out is to use the instrument under the conditions in which it will be employed.

Good clinical laboratories have a recording spectrophotometer. Instruments of this type can be used for identification of compounds by the location of wavelengths of maximum and minimum absorbance. Also, it has the capacity to resolve multicomponent solutions by the use of absorptivities, illustrated by the above example. A very practical application of these instruments is to evaluate the accuracy and precision of assays using absorbance measurements under the best conditions. The latter is very important as a component of a good quality control program.

Many assay methods that are used in clinical laboratories have not been thoroughly investigated. Many procedures have been developed using filter photometers that utilize measurements at wavelengths which are not optimum. A recording spectrophotometer can provide the spectra of the reagent blank, sample blank, sample, and sample plus substances known to or suspected to interfere. From such data a wavelength can be selected which provides the best absorbance measurements. Some assays allow for measurements over a wide area, whereas others require a very narrow portion of the spectrum.

OTHER TYPES OF PHOTOMETRY

TURBIDIMETRY AND NEPHELOMETRY

So far, absorbance has been concerned with radiant energy absorbance by the particular molecular structure. Turbid solutions behave in a manner similar to colored solutions. The particles absorb, reflect, and refract light in proportion to the number and size of the particles. Instruments used for absorbance measurements can be used for turbidity measurements and when so used are sometimes called turbidimeters (Fig. 18-20).

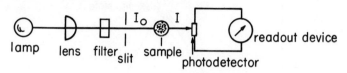

Fig. 18-20. Turbidimeter.

Nephelometry is a technique for measuring the light scattered by turbid solutions (Fig. 18-21). The photodetector is placed at a

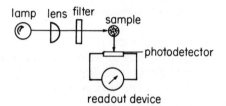

Fig. 18-21. Nephelometer.

right angle to the source so that only reflected and refracted light is detected. In general, nephelometry is a little more sensitive than turbidimetry.

Both techniques often provide a means of measuring a substance that can be precipitated or flocculated without the need to isolate the material. The short wavelengths scatter more and are generally used, if possible. The amount of light scatter is related to particle size, and it is therefore important that conditions be established so that consistent particle size can be obtained. This is difficult to accomplish in practice.

The Beer-Lambert relationships **350** (b) do not apply in turbidimetry and nephelometry. However, in some instances there is a linear relationship between concentration and turbidimetric and nephelometric values.

There are so many problems associated with these two techniques, it is generally better to avoid them if at all possible.

FLUOROMETRY

Many substances in solution will fluoresce when radiated with UV light (solution fluorescence). When these substances absorb ultraviolet light they are raised to a higher energy level, often called an excited state. This is illustrated as a time plot in Figure 18-22.

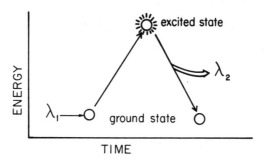

Fig. 18-22. Light sorption and emission and energy changes in fluorescence.

When molecules in the ground or lower energy state are subjected to ultraviolet light (λ_1), they absorb energy to reach the excited state. The molecules are unstable at this higher energy level and quickly return to the ground state. The energy that is lost as they return to the ground state is emitted as radiation of light at a longer wavelength (λ_2). Notice in Figure 18-22 that the molecules are excited by shorter wavelength light (λ_1) and fluoresce at a longer wavelength (λ_2). Recall that the energy content of radiation is inversely proportional to wavelength. Thus, the light used to activate the molecules has more energy than that emitted as fluorescence. The difference in energy represents losses in the process, mainly as heat.

Phosphorescence is the same as fluorescence, except that the emission lasts for some time after activation has stopped. Fluorescence stops when activation stops.

A fluorometer has the same basic structure as a nephelometer, with a few accessories (compare Fig. 18-21 to Fig. 18-23).

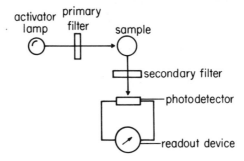

Fig. 18-23. Fluorometer.

The activator lamp is a source of ultraviolet light for exciting the sample. The primary filter or other wavelength-selecting equipment (grating, prism) is used to isolate the portion of the spectrum desired to excite the sample molecules. A secondary filter (grating, prism) isolates the fluorescent light from the activator light and also the desired portion of the fluorescent light. The photodetector and readout device are the same as those described previously.

Fluorescence is emitted light, and the Beer-Lambert law does not apply. Equations can be developed that relate concentration of fluorescent substances to the amount of fluorescence, but so many factors influence fluorescence that it is generally better to use several standard solutions and plot fluorescence versus concentration each time an analysis is performed. The plot can be used to determine the concentration values for the samples.

Fluorescence measurements are quite different from absorbance. For the latter, the ratio of I_0 to I is measured. In fluorometry only fluorescent light from the sample is considered. The instrument is adjusted to zero fluorescence by using a solid black object ("blank") the same size and shape as the cuvet. Theoretically, any fluorescent light from the sample could be measured. The sensitivity of the photodetector, of course, determines the amount of fluorescence required for valid measurements. Most fluorometric procedures are 1,000 to 10,000 times more sensitive than absorbance methods.

Some compounds fluoresce without any manipulation. Others can be measured fluorometrically only after a chemical reaction, such as oxidation, reduction, or after degradation or derivatization.

The primary activation wavelength selection often needs to be good enough to provide a very narrow band so that only the desired compound is activated. When working with biologic material one is dealing with a mixture of many fluorescing compounds, and if the sample were irradiated with a wide band many compounds would fluoresce.

Similar reasoning can be applied to the secondary filter or other device for wavelength selection. The more selective the primary and secondary wavelength selectors are, the more specific the analyses can be.

EMISSION FLAME PHOTOMETRY

The first three alkali metals—sodium, potassium, and lithium—can be activated to emit light in much the same way as molecules are activated to fluoresce. When these metals are heated the atoms are raised to higher energy levels, excited states, and as they return to ground levels they emit light. Under controlled conditions the isolation and measurement of the emitted light allows quantitation of these elements.

A very simple flame photometer is represented schematically in Figure 18-24. In this system the sample containing the element

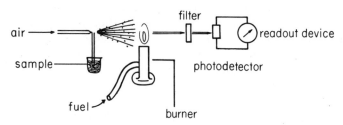

Fig. 18-24. Simple flame photometer.

to be assayed is aspirated and nebulized employing the Venturi principle. The sample aerosol is directed into the flame, and the heat energy of the flame excites the atoms to emit light. The desired emission is isolated by the filter and measured by the photodetector-readout device combination.

In reality flame photometers are much more complex. If the nebulizer is the type pictured, only a small portion of the aerosol is volatilized by the flame, and a mechanism is required to collect and dispose of the excess. This is usually accomplished by a glass nebulizer chamber equipped with a drain. Some flame photometers employ "total consumption" systems whereby only a small amount of sample is introduced directly into the flame. Fuel pressure and the heat of the flame accomplish the nebulization, so these total consumption systems do not require nebulizer chambers.

The purpose of nebulization is to introduce uniformly small amounts of the sample solution into the flame. The solvent is vaporized leaving minute particles of dry materials. The organic substances are destroyed by combustion, leaving inorganic salt residues. The heat of the flame converts the elements to be measured to neutral atoms. The thermal energy of the flame excites the atoms, which emit light as they return to the ground state.

Spectra isolation of the emitted light is required so that only the light from the element being assayed is measured. Thus, flame photometers usually have the optical system protected from extraneous light by appropriate light shields.

When a system as diagrammed in Figure 18-24 is used for quantitation, measurements of the light emissions of standard solutions are compared with solutions of unknown concentrations. This type of instrument, known as a "direct reading" flame photometer, is not very practical because of several sources of error. Changes in air pressure or flow due to variation in pressure or obstruction can be a source of error. For example, if a decrease in air pressure or flow occurs following aspiration of the standard solutions, less unknown would be aspirated and thus less light emitted. The result in this case would indicate a smaller concentration than was actually present in the sample. For reliable results the air pressure and flow must be constant.

A similar problem can occur with nebulization because the sample tube in the chamber has a very small internal diameter and fibrin and other particulate matter can obstruct sample flow. If the light emitted from the standard solutions is measured with an unobstructed nebulizer and an obstruction occurs during aspiration of an unknown, less sample will be aspirated, again resulting in an erroneously low result.

The temperature of the flame is directly proportional to light emission. Therefore, variations in the fuel supply will have profound effects on the results obtained. Many flame photometers utilize "city gas" and pressure in the gas main usually follows temperature fluctuations during rapid weather changes.

The reverse of the above two processes, increases in pressure or flow of the air or fuel, can also occur. This leads to erroneously high results.

The difficulties can be largely overcome by using the "internal standard" approach. This type of instrumentation requires an additional photodetection device and is represented in Figure 18-25. Here there are two additional photodetecting devices, but only two

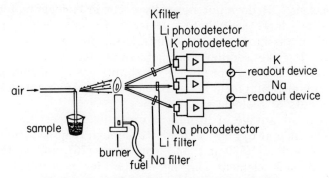

Fig. 18-25. Internal standard flame photometer.

are required to measure one element — that is, only the lithium and sodium units would be required to measure sodium. In practice it is generally desirable to measure both sodium and potassium simultaneously on the same sample, and the equipment represented can accomplish this with three photodetecting units.

To use the internal standard technique it is necessary to have a constant concentration of lithium in the samples. This is easily accomplished by diluting all the samples with a large volume of a lithium solution. The dilution is usually of the order of 200 times. The equipment is adjusted for zero sodium and potassium concentration by aspirating only the lithium diluting solution. The light emitted by the lithium is constant, and the variables are the emissions of sodium and potassium. Quantitation is accomplished by comparing the ratios of sodium and potassium emissions to that of lithium.

The advantages of the internal standard are obvious when the usual difficulties of flame photometry are considered. Because absolute emissions are not measured, only the ratios, a decrease in air pressure has little effect. The ratios Na/Li and K/Li are the same regardless of the amount of sample aspirated. Similarly, if the flame temperature changes and thus affects the magnitude of the emissions, the effect will be minimal because the emission of all three elements will be changed in the same direction.

This is not meant to imply that the use of an internal standard solves all the problems associated with flame photometry. Indeed it does not, but it does help. Flame photometers are notoriously "fussy" instruments, and good, reliable performance requires considerable attention. Cleanliness is perhaps most important. The nebulizer system, burner, and chimney must be cleaned at regular intervals. The air and fuel should be supplied at constant pressures and flow rates.

Lithium is commonly used as an internal standard for the following reasons. It does not normally occur in biologic materials at concentrations large enough to interfere. Its salts, which are free of sodium, potassium, and other interfering materials, are readily available. It, like sodium and potassium, has a good emission yield — that is, the heat produced by simple gas-air flames is sufficient to produce ample light emission for easy measurement with small amounts of biologic samples.

A great advantage of lithium is its emission spectrum. It has a strong emission at 671 nm, which is far enough removed from useful emission of sodium (589 nm) and potassium (767 nm) so that the emission of each element can be isolated for measurement with simple interference filters.

Flame photometry can be used to measure calcium and magnesium, but much more elaborate instrumentation including a very hot flame is required. These elements can be better assayed by other techniques, particularly atomic absorption spectrophotometry.

ABSORPTION FLAME PHOTOMETRY

Absorption flame photometry or atomic absorption spectrophotometry utilizes principles that have been known for many years but have been employed only recently for analyses. The basic principle is that atoms at the ground state absorb light of the same wavelengths they would emit if they were excited. Flame emission photometry is generally applicable only for a few elements (such as sodium, potassium, and lithium) because most elements are difficult to excite in an ordinary flame. The absorption of light at the emission wavelength is a different matter. Many elements show sufficient absorption to enable quantitation.

The principle of the technique can probably be better appreciated after a consideration of the instrument in Figure 18-26.

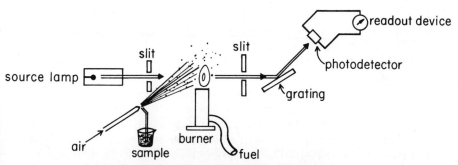

Fig. 18-26. Atomic absorption photometer.

The source lamp has a cathode made of the element to be measured. When the cathode is heated by an electric current the element becomes excited and emits light. The sample is nebulized and passed into a flame much like in flame photometry. The atoms in the flame absorb the light from the source in direct proportion to their concentration. A diffraction grating or similar device for spectral isolation is required so that only the desired wavelength is measured and other light from the flame is rejected. The instrument is adjusted to zero absorbance by aspirating a solution containing none of the element to be measured. When a sample is aspirated the decrease in light reaching the photodetector is a measure of light absorbance and, in turn, is directly related to the concentration of the element to be measured.

Actual atomic absorption spectrophotometers are much more complex than shown in Figure 18-26. Here there are no means for distinguishing between the light from the source lamp and that from emission by the element in the flame. Some instruments compensate for light emission by using a modulated (chopped) source and an amplifier synchronized with the modulation. Emission light would then appear as a DC signal and would not be amplified, whereas the source light, being modulated, would be amplified.

The burners used in most instruments are long and narrow with the source light directed through the long axis of the flame to increase the "depth" of the sample. In addition, most burners are the "premix" type in which the sample is nebulized into the fuel mixture and the resulting aerosol is burned. Several different fuels are commonly used, but acetylene-air is very common.

Some commercially available absorption flame photometers are of the direct reading type shown in Figure 18-26, some have a split beam, and others have resonant detectors. The entire field of atomic absorption spectrophotometry is so new that it is still in a state of flux. As a result, there is no generally accepted technique or instrumentation, and further discussion of these topics is premature.

One other aspect of this analytic approach, flame fluorescence, probably deserves mentioning. It is analogous to fluorescence as previously described. When irradiated with exciting light certain molecules dispersed in solution emit light (solution fluorescence). Similarly, atoms dispersed in a flame when irradiated with the proper wavelength absorb light and are excited to emit light. To measure this flame fluorescence the instrumentation is similar to that shown in Figure 18-26 except that the photodetector is at a right angle to the source. This is also analogous to solution fluorometers. Flame fluorescence is not in general use in clinical laboratories at present but does offer promise of future application.

The advantage of absorption flame photometry is its sensitivity. As indicated previously, flame emission photometry is generally practical only for the measurement of sodium, potassium, and lithium. The reason most metals cannot be assayed by flame emission techniques is that they are very difficult to excite with ordinary flame temperatures. The absorption of light characteristic of the element is quite another thing. Apparently most elements, including gold and silver, absorb their characteristic wavelengths readily. Thus, meaningful results can be obtained. Furthermore, by careful selection and good isolation of the wavelengths measured the technique can be very specific. Atomic absorption spectrophotometry is commonly used in clinical laboratories to measure calcium, magnesium, copper, lead, zinc, iron, mercury, and thallium in biologic materials.

CHAPTER 19

QUALITY CONTROL

The quality of manufactured items can be controlled relatively easily. The manufacturer of pencils knows before starting production exactly what he wishes to produce—the length, diameter, weight, color, and hardness of the lead. During the assembly process these criteria can be examined and the necessary adjustments made to ensure a quality product. Pencils that are 4 feet long or have erasers on both ends deviate from the desired product, and the defective operation is easily identifiable.

The situation is considerably more difficult in clinical chemistry where the product, the analytic value, is unknown. There are some practical limits for acceptable values, but still the value is largely unknown. For example, a total serum protein value of 50 g/dl is untenable but a value between 2 and 12 g/dl is possible. The range of possible values is even greater in some cases: glucose 25 to 1,000 mg/dl, and blood urea nitrogen 5 to 200 mg/dl.

Other features contribute to the problem. Most methods used in clinical chemistry lack specificity, and when analyzing biologic materials, which are quite variable and very heterogeneous, this lack of specificity can cause considerable difficulty. The assay of a urine sample for glucose by two methods will show differences due to method specificity. With a method that employs alkaline ferricyanide, values of 50 to 500 mg/dl may be obtained, whereas using o-toluidine in acetic acid as the color reagent, values of 15 to 150 mg/dl would be obtained.

In certain procedures the presence of bilirubin in the serum can interfere with the analysis of protein and cholesterol because bilirubin absorbs light at the same wavelengths as those used in the assays.

This interference can be quite variable because the concentration of bilirubin can range between 1 and 20 mg/dl.

The aspects of specificity and interference are applicable to most clinical chemistry assays, but these examples are sufficient to indicate the type and magnitude. The pencil manufacturer can readily and easily know if he has lead in his pencil, but bioanalysts have more difficult problems.

One approach to this problem would be to use more specific methods — e.g., the utilization of specific enzymes as reagents. However, the problems of enzyme activities and endogenous inhibitors plague these endeavors. The multitude of methods usually available for the analysis of any one substance attest to the fact that there are few really reliable methods.

The importance of valid laboratory results is obvious. Decisions regarding drug therapy, surgery, and length of hospital confinement are frequently made on the basis of laboratory results. Valid results are possible only with good techniques and good assay procedures. Both of these can be regulated with meaningful quality control. Before the subject of quality control is explored further, a slight regression to data manipulation and statistics is mandatory.

LABORATORY DATA AND STATISTICS

SIGNIFICANT FIGURES

No healthy young man would deny that 38-24-36 is a rather significant figure, but what about 79.33? A general rule is that there should be no more significant figures in an answer than in any of the original data. Suppose one wishes to divide 10 cm by 30.0

386
$$\frac{10 \text{ cm}}{30.0} = 0.333333\ldots \text{ cm}$$

Mathematically, the answer has an infinite number of digits. The original value of 10 cm had 2 significant digits, so the answer should also have 2 and be recorded as 0.33 cm. Some examples showing the number of significant figures follow.

387
        79   2 significant figures
      6348.39   6 significant figures
      0.000173   3 significant figures

Along with the concept of significant figures there must also be an appreciation of the significance of numbers. To a mathematician a number is a number, but to a scientist a number also has a significance. For example, the number 10 means any value between 9.5 and 10.5; 10.5 means values between 10.45 and 10.55, and so on. Analytic results cannot have more significance or significant figures than the least accurate analytic value. Suppose a titration were performed with 0.10965 N NaOH and a 10-ml serologic pipet to provide a titer of 7.6 ml. The milliequivalents delivered by the pipet would be

388
$$\text{ml} \times N = \text{meq}$$
$$7.6 \times 0.10965 = 0.83334$$

The mathematical value of 0.83334 would have significance of only 0.83 meq.

ROUNDING NUMBERS

In **386** the string of threes was chopped off to leave just two numbers — that is, the number was rounded off. Manipulations involving multiplication and division, particularly when performed using calculators, provide more numbers than can be honestly used and the numbers must be rounded off. The general rules are: if the figure to the right of the last digit to be retained is less than five, the number retained is left unchanged; if the number is greater than five, the digit to be retained is increased by one.

389
7.684 ⟶ 7.68
7.686 ⟶ 7.69

If the number to the right of the number to be retained is five, a "fielder's choice" ensues. As an aid in decision-making in this instance, another rule is commonly invoked: when the number to the left of 5 is odd, it is increased by one, but if it is even, it is left unchanged.

390
0.0535 ⟶ 0.054
0.0545 ⟶ 0.054

Some care must be exercised in rounding numbers; it should not be done until the final number is obtained. Suppose that the original numbers as obtained in **390** were to be multiplied by 66 to obtain an answer.

391
$0.0535 \times 66 = 3.5310$ ⟶ 3.5
$0.0545 \times 66 = 3.5970$ ⟶ 3.6

Certainly, different answers are obtained. If the rounded number is used

392
$0.054 \times 66 = 3.564$ ⟶ 3.6

the larger value is obtained in this instance. The general rule is to "carry" more figures than can be used for the final number and round the number as the last operation.

DATA DISTRIBUTION

If one were to analyze a serum sample five times for urea nitrogen, the following results might be obtained.

393

	1	20.1 mg/dl
	2	19.4 mg/dl
	3	21.3 mg/dl
	4	24.5 mg/dl
	5	18.9 mg/dl

When the results of replicate analyses agree pretty well, it is common practice to use the mean as the analytic result. A review of the data in **393** reveals that value number 4 seems a little out of line as compared with the rest. Is value number 4 within the acceptable limits for the method or does it represent a deviant value which should be disregarded? Before a decision can be made, the acceptable limits for the method must be established by determining some features of the analytic values using statistical techniques.

Frequency Distributions

If the weights of 8,178 male college seniors were divided into 10-pound weight intervals and tabulated by increasing weights, the data might appear as in Table 19-1.

TABLE 19-1. FREQUENCY DISTRIBUTION OF WEIGHTS OF MALE COLLEGE STUDENTS

Weight interval	Number	Weight interval	Number
60-69	0	190-199	730
70-79	3	200-209	546
80-89	15	210-219	398
90-99	38	220-229	236
100-109	76	230-239	138
110-119	119	240-249	85
120-129	284	250-259	44
130-139	564	260-269	12
140-149	860	270-279	4
150-159	1029	280-289	7
160-169	1071	290-299	1
170-179	1015	300-309	3
180-189	882	310-319	0

Average = 172.2 ⟶ 172

A frequency histogram constructed with these data would appear as Figure 19-1.

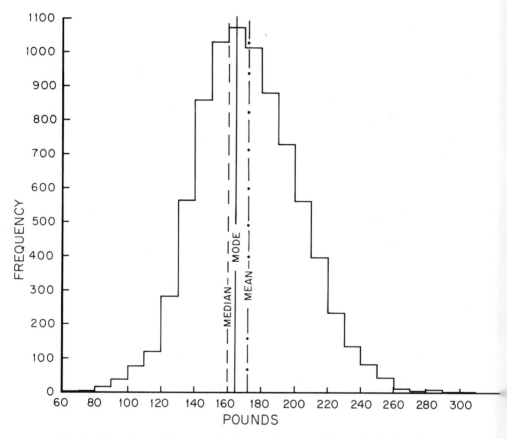

Fig. 19-1. Frequency histogram of the weights of male senior college students (see Table 19-1).

The arithmetic average or mean is defined as the sum of the values divided by the number of values. This definition can be expressed using common symbolism.

$\bar{X}$ = mean

$X_1, X_2, X_3, \ldots X_N$ = individual values

N = number of values

$$\bar{X} = \frac{X_1 + X_2 + X_3 + \ldots + X_N}{N}$$

For further simplification the summation symbol, Σ, can be used.

395
$$\bar{X} = \frac{\Sigma X}{N}$$

The mean represents the arithmetic central value for the data. The mean for the male college seniors is 172 pounds. There are two other measures of central values: the median and the mode. The median value is the midpoint in the distribution — that is, one-half of all the values are less than the median and one-half are greater. The median for the data collected on the college men is 160 pounds. The mode is the value that occurs most frequently for a distribution, or is the midpoint of the interval with the greatest number of cases. For the collegians considered here the mode would be 164.5 pounds, which would be rounded to 164 pounds.

The mean, median, and mode from the data of Table 19-1 are noted in Figure 19-1. Notice that although all three are measures of central tendency, they are different for these data. To better illustrate the uses of these three measures consider the following. After a class of 10 students of clinical chemistry had finished their course, they decided to purchase a gift for their instructor to express their appreciation for his efforts. The bookkeeping appeared as follows:

396

Student	Contribution
GAW	$ 0.05
PA	0.09
RH	0.25
MS	0.15
HB	20.00*
WM	0.04
BP	0.12
ES	0.27
BD	0.01†
MG	0.19

* Contribution by the student who earned an A
† Contribution by the student who earned an F

The mean contribution of each class member is $ 2.12, but the median is $0.135 or $0.14. Obviously, the median is a better indication of the students individual contribution than is the mean.

Normal, Bimodal, and Skewed Distribution

If data are collected on a large number of random events and plotted using very small intervals, a smooth curve will result by

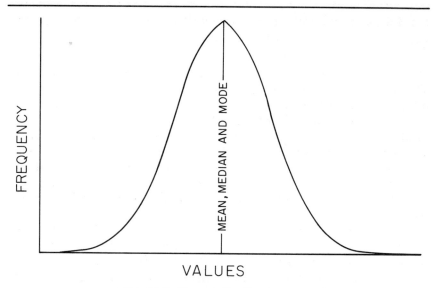

Fig. 19-2. Normal distribution curve.

connecting the midpoints of the intervals (Fig. 19-2). This is a normal distribution curve and has some characteristics that should be noted. The mean and median are in the center of the curve and have the greatest frequency. Thus the mean, median, and mode are identical. It can be constructed only from data that have random variation from the mean value — that is, there must be an equal number of values greater than and less than the mean. The curve is asymptotic in that it never reaches the abscissa.

Data to construct a normal curve can be gathered from simple experiments. Consider a coin toss game using eight coins. The results will be random because each coin has the same probability of landing as a "head" or "tail." To gather the data the eight coins would be tossed 1,000 times. Table 19-2 shows the ideal distribution

TABLE 19-2. FREQUENCY DISTRIBUTION FOR COIN TOSS

Heads	Tails	Frequency
8	0	4
7	1	31
6	2	109
5	3	219
4	4	273
3	5	219
2	6	109
1	7	31
0	8	4

of heads and tails for the coin toss game. A plot of these data is shown in Figure 19-3.

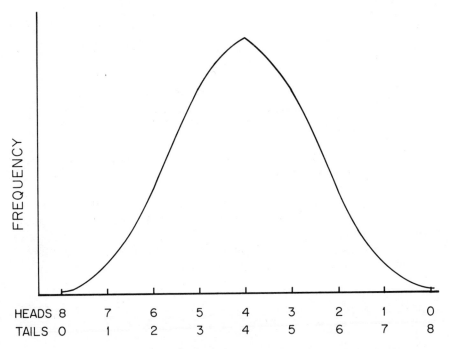

Fig. 19-3. Frequency distribution curve for coin toss (see Table 19-2).

The data from the coin toss game indicate that if the game is to be played for money, then the best bet would be four heads and four tails, because the combination occurs most frequently. Furthermore, if one were to bet on eight heads or eight tails, the odds should be 8 : 1,000.

Indeed, statistics and probability science got their starts in the eighteenth century at the request of gamblers who needed to know how to make more intelligent bets.

In actual practice the data collected seldom, if ever, will describe a normal distribution curve. However, if the data approximate a normal distribution, the properties of the normal distribution curve can be used to provide an evaluation of the data, with little error.

It was indicated above that deviations from the mean must be random to have a normal distribution. Consider another weight-data-gathering adventure, this time using varsity football players (Fig. 19-4).

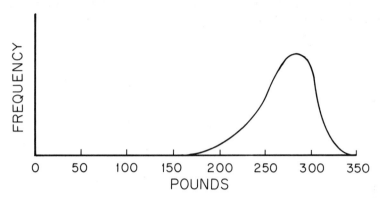

Fig. 19-4. Frequency distribution curve of the weight of varsity football players showing negative skewness.

The weights in this example are not uniformly distributed but rather are accumulated to the right. Such a distribution is said to be negatively skewed.

Next consider the spendable money per month of a large number of college students as shown in Figure 19-5.

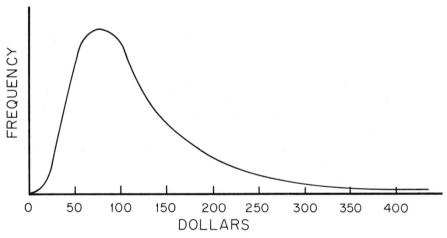

Fig. 19-5. Frequency distribution of college students' spendable income showing positive skewness.

Thanks to a few rich students, the curve is skewed. This type of distribution is called positive skewness.

Distribution can be multimodal. An example of a bimodal distribution might be obtained from data on the weights of both female and male college seniors (Fig. 19-6).

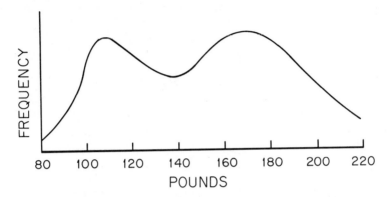

Fig. 19-6. Bimodal frequency distribution of the weights of both male and female college students.

Here the girls would show greatest frequency around 110 pounds and the boys, as before, would be clustered around 165 pounds. Distributions of this type indicate that the data are from two different populations and each set should be considered separately.

Skewed, bimodal, and types of distributions other than the so-called normal require special treatment that will not be considered here. If the data describe a distribution that is symmetrical with respect to the mean, however, some simple calculations can be made to obtain useful information about the data.

Standard Deviation

The letter " s " is a symbol commonly used for standard deviation. The area representing 99% of the total area of a normal distribution curve can be divided on the abscissa into six equal lengths using vertical lines marked ±1s, 2s, and 3s (Fig. 19-7).

The area between the lines labeled − 1s and + 1s represents 68% of the total area; − 2s to + 2s equals 95%; and − 3s to + 3s is 99% of the total area. Considering the areas in another way, 68% of all the values fall between ±1s; 95% between ±2s and 99% between ±3s. In a normal distribution 68% of all the values would fall between ±1s and only 0.5% of the values would be greater than + 3s or less than − 3s. The importance of s is obvious then, because it provides information about the distribution of data.

266   Clinical Biochemistry

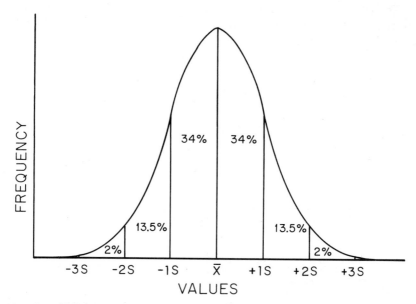

Fig. 19-7. Normal distribution curve showing the areas included in 1, 2, and 3 standard deviations.

Standard deviation can be calculated by the use of two different formulas. The formula used depends upon the type of data to be evaluated and whether a calculator is available. One formula is

397

X = individual values

$\bar{X}$ = mean

d = $(X - \bar{X})$; the difference of each individual value from the mean

Σd^2 = sum of d^2

N = number of values

$$s = \sqrt{\frac{\Sigma d^2}{N-1}}$$

The calculation of s using a series of cholesterol values will help to clarify the formula (Table 19-3).

For these data, $\bar{X} \pm 1s$ is 181 to 215, $\bar{X} \pm 2s$ is 164 to 232, and $\bar{X} \pm 3s$ is 147 to 249 mg/dl. A review of the data shows that 18 or 72% of the values lie within $\bar{X} \pm 1s$; 24 or 96% are between $\bar{X} \pm 2s$; and 100% are within $\bar{X} \pm 3s$. This distribution is different than the normal distribution and is typical of that obtained when using experimental values.

TABLE 19-3. CHOLESTEROL VALUES AND CALCULATION OF STANDARD DEVIATION

Sample No.	Cholesterol (mg/dl) X*	d†	d²‡
1	198	0	0
2	194	4	16
3	208	10	100
4	232	34	1156
5	219	21	441
6	188	10	100
7	188	10	100
8	177	21	441
9	190	8	64
10	200	2	4
11	197	1	1
12	154	44	1936
13	199	1	1
14	191	7	49
15	194	4	16
16	189	9	81
17	188	10	100
18	188	10	100
19	180	18	324
20	213	15	225
21	227	29	841
22	222	24	576
23	211	13	169
24	214	16	256
25	199	1	1

$\Sigma X = 4,960 \qquad \Sigma d^2 = 7,098 \qquad \bar{X} = 4,960/25 = 198.4 \longrightarrow 198$

$$s = \sqrt{\frac{\Sigma d^2}{N-1}} = \sqrt{\frac{7,089}{25-1}} = \sqrt{295.7} = 17.2 \longrightarrow 17$$

* Obtained by analyzing same sample 25 times.
† Difference of each individual value from the mean.
‡ Square of the difference of each value from the mean.

Another formula commonly used to calculate standard deviation is

398

X = individual values
ΣX^2 = sum of the values squared
$(\Sigma X)^2$ = square of the sum of the values

$$s = \sqrt{\frac{\Sigma X^2 - \frac{(\Sigma X)^2}{N}}{N-1}}$$

The same 25 cholesterol values used in Table 19-3 are used again to illustrate this equation (Table 19-4). The standard deviation value is the same as for the preceding calculation.

TABLE 19-4. CHOLESTEROL VALUES AND CALCULATIONS OF STANDARD DEVIATION

Sample No	Cholesterol (mg/dl) X	X^2
1	198	39,204
2	194	37,636
3	208	43,264
4	232	53,824
5	219	47,961
6	188	35,344
7	188	35,344
8	177	31,329
9	190	36,100
10	200	40,000
11	197	38,809
12	154	23,716
13	199	39,601
14	191	36,481
15	194	37,636
16	189	35,721
17	188	35,344
18	188	35,344
19	180	32,400
20	213	45,369
21	227	51,529
22	222	49,284
23	211	44,521
24	214	45,796
25	199	39,601
	$\Sigma X = 4,960$	$\Sigma X^2 = 991,158$

$\overline{X} = 198$

$(\Sigma X)^2 = 24,601,600$

$$s = \sqrt{\frac{\Sigma X^2 - \frac{(\Sigma X)^2}{N}}{N - 1}}$$

$$s = \sqrt{\frac{991,158 - \frac{24,601,600}{25}}{25 - 1}}$$

$$s = \sqrt{\frac{991,158 - 984,064}{24}} = \sqrt{\frac{7,094}{24}} = \sqrt{295.58}$$

$s = 17.4 \longrightarrow 17$

Now that all these mathematical manipulations have been presented, what does it all mean? These elementary statistical techniques are applicable to laboratory quality control programs. The question of the validity of value number four in **393** can be answered with these techniques.

QUALITY CONTROL TECHNIQUES

ACCURACY AND PRECISION

The first aspects of an analytic procedure to be considered are the accuracy and the precision. The term accuracy means how close the analytic value is to the accepted value. The problem arises in establishing accepted values. Recourse is usually made to "reference methods" to establish procedure accuracy. Precision is a term used to describe the variation within a series of analytic results. If a single sample of serum were analyzed for potassium and the following values obtained

399

1	4.1 meq/liter
2	4.2 meq/liter
3	4.1 meq/liter
4	4.1 meq/liter
5	4.2 meq/liter

the conclusion would be that the analyses show good precision. However, if by other accepted techniques it was found that the potassium concentration was 6.2 meq/liter, then the procedure would have little accuracy. Analytic methods can have good accuracy and good precision, or good accuracy and poor precision, or vice versa.

The accuracy of a procedure depends primarily on two factors: the specificity of the method and the technique employed. If a method is not very specific and is subject to many interferences, it will not provide reliable or accurate results. Technique cannot be overemphasized. It is obvious that the best method in the hands of a poor analyst will not yield accurate results. Precision is a measure of the reproducibility of a method and is largely a function of instrument response and technique. In clinical chemistry inaccurate methods frequently are employed simply because better methods are not available.

Assuming that the most accurate methods and the best instrumentation available are used, then the emphasis is placed on the precision or reproducibility. To establish the reproducibility of a method, replicate analyses are performed on the same sample under the same conditions. A minimum of 20 analyses can be used, but it is preferable to use 40 or more. As the number of values increases, the magnitude of s decreases. The analytic values are then examined to determine their distribution. This can be done by plotting a frequency histogram or, with experience, by inspection of the data grouped as in Table 19-1. If the distribution approximates a normal distribution and is not multimodal or skewed, then s is calculated. If all the factors which contribute to variation in the results are random then $\bar{X} \pm 3s$ should cover 99% of all the results. The data in Tables 19-3 and 19-4 show that at the 198 mg/dl level a variation of $\pm 3s$ gives a range of 147 to 249 mg/dl. This is obviously not a very precise method because the data indicate that if a sample actually contained 198 mg/dl there is the possiblity that it could be reported to have any concentration between 150 and 250 mg/dl.

Factors That Influence Precision

There are many identifiable factors which influence precision or reproducibility. There is variation in the precision when parts of the data are collected from analyses done during different days (day-to-day variation); there also is variation between individuals performing the analyses, the types of samples (aqueous standards versus serum, etc.), equipment, reagents, the time of the day, and many others. To obtain meaningful measures of precision it is therefore necessary to include all these factors. If an analysis is to be performed by 10 different people, then the data should be obtained from all 10. Also, if the analyses are made by technical personnel during three different work periods per day, the data should be gathered from each work period and from each individual performing the assay – i.e., the data for method reproducibility should be collected and analyzed under the same conditions and at the same times as biologic specimens.

In addition to knowing the precision under usual laboratory conditions, it is also important to know the precision under the best possible conditions. To collect data for the latter, technically competent personnel should perform an adequate number of replicate assays, preferably during a single day or less, with a minimum of distraction or other interference. An indication of precision under the best conditions provides a goal to strive for under the usual conditions. A great difference in method precision when determined under both the best and usual laboratory conditions suggests that the usual conditions should be improved.

If care is not exercised human bias can intervene. It is well established that an individual is apt to produce an analytic result nearer the accepted value when the value is known than when it is unknown. To avoid this source of bias, samples used to collect reproducibility information can be disguised and presented to the analyst as unknowns.

A factor which greatly affects the magnitude of variation with respect to the mean is the value of the mean. Consider the measurement of serum bilirubin. If s for this assay were 0.1 mg/dl and the mean were 0.5 mg/dl, then the precision would appear to be poor. If, however, the mean were 18 mg/dl, then the precision would appear to be reasonably good. This variation with respect to the mean is sometimes reported as the coefficient of variation (CV).

400

$$CV = \frac{s}{\bar{X}} \times 100$$

For the first example of bilirubin analysis the CV would be

$$CV = \frac{0.1}{0.5} \times 100 = 20\%$$

but for the second example it would be

$$CV = \frac{0.1}{18} \times 100 = 0.6\%$$

It is therefore obvious that values for s and CV are meaningless unless the magnitude of $\bar{X}$ is known.

The range of the values used to calculate precision is also important. If the $\bar{X} \pm 3s$ excludes an appreciable number of values, it suggests that there is more than random variation or that the distribution is not normal.

ACCEPTABLE LIMITS

When s, $\bar{X}$, and the range of the analytic values have been found, then acceptable limits for a procedure can be established. Most clinical chemistry procedures will have 95% or more of the values within $\pm 2s$ and $\pm 3s$. To return to the blood urea nitrogen example in **393**, if the s for blood urea nitrogen assays at the 20 mg/dl level were ± 0.8 mg/dl and acceptable limits were established at $\pm 2.3s$, then the acceptable variation would be $\pm 2.3 \times 0.8 = \pm 1.8$ mg/dl. The average for the results in **393** is 19.9. This would provide an acceptable range of 18.1 to 21.7 mg/dl and would exclude value number 4, 24.5 mg/dl.

Frequently it is necessary to know if one set of data is the same as or different from another set. For instance, an assay technique is changed in some way, and the question arises as to whether the analytic values obtained are the same or different. If the means and standard deviation are similar it does not mean that the results are the same. The evaluation of two different sets of data is often very complex and requires more advanced concepts and calculations than will be presented here.

To summarize the process of establishing an acceptable range of values, it is first necessary to collect unbiased analytic data under the usual assay conditions. The distribution is evaluated, and if it approximates a normal distribution the $\bar{X}$ and s are determined. Then the range of values that encompasses 95% or more of the analytic values is found, and this range is then the acceptable range.

STANDARD SOLUTIONS AND CONTROL SAMPLES

A standard solution has a known concentration and purity, and is required as a reference for quantitation. Standard solutions can also serve as a part of quality control; if all the assay conditions remain the same, the absorbance, fluorescence, titer, and so on for the standard solutions should be the same from one set of analyses to the next. In this respect the values obtained from the standard solution serve as a type of control.

A standard solution is usually a single substance dissolved in a suitable solvent and is generally quite different than biologic specimens. In addition, standard solutions are apt to behave differently in an analytic procedure. For example, deproteinizing reagents may be added to standard solutions in a procedure, but if they contain no protein, there obviously will be no precipitation.

The best way to monitor an analytic procedure is to use one or more control samples in each assay. A control sample has a composition that closely mimics the biologic material being assayed. If the analysis is performed on serum then a serum control sample is employed, and if it is a urine analysis, then a urine control sample is used, and so on.

Control samples are available commercially or they can be obtained by pooling the appropriate specimens. The concentrations of the constituents of interest must be determined with valid assay procedures and under the conditions which the control sample will be used. Because of the variety of analytic techniques and laboratory conditions it is hazardous to accept values obtained from other laboratories and possibly with other procedures.

A control sample is assayed with each analysis. If its analytic value falls out of the acceptable range then all of the analytic values are suspect. Usually the use of the control sample will not indicate what is wrong with the analytic process, but simply that something is wrong. It could be that the assay was appropriate for every sample but the control sample, but this has to be verified. If the control sample result is out of range then analysis of only the control sample can be repeated; this avoids the use of sometimes hard-to-get biologic specimens. Many laboratory workers analyze control samples first and do not jeopardize patient material until they are sure the analytic system is in control.

Added assurance is provided by employing two or more different control samples, one in the normal range and the others with abnormal values.

More information can be obtained from control samples if the analytic data are accumulated. One way of using these data is to plot the average value for each of the control samples every day. This is seen in Figure 19-8. This shows a plot for a serum sodium

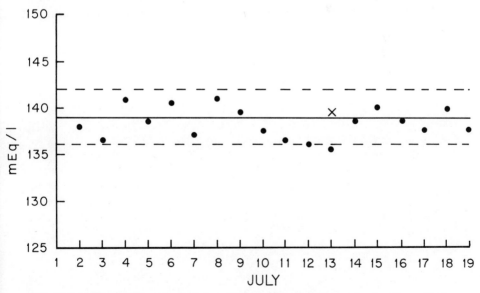

Fig. 19-8. Daily quality control plot. *Solid line* represents the mean for the control sample, and *dashed lines* show the acceptable limits.

control sample on a daily basis. The data plotted from July 1 to July 8 shows the pattern expected; the results are random above and below the mean. On July 8 or 9 trouble develops. The control values continue to decline until July 13 when a value obtained is too low to be acceptable. The pattern suggests that the concentrations of the aqueous standard solutions are too high. Perhaps someone was leaving the tops off the bottles and the standard solutions were concentrated by evaporation. Regardless, new standard solutions were prepared and reanalysis of the control serum yielded 139.5 meq/liter (X on plot), and the rest of the data thereafter remain in control.

There are several other techniques for following the time course of control sample values. The advantage of this approach is that frequently there will be trends which predict that the analytic procedure is going out of control, and the direction of the trend may give some indication of the problem.

INDEX

absorbance, 226
  photometers, 227
  photometry, 223
absorption flame photometry, 253
absorption, intestinal, 97
accuracy, 269
acetoacetate, 121
acetoacetyl CoA, 118, 121, 122
acetone, 122
acetic acid titration, 89
acetyl coenzyme A, 116, 121, 146
achlorhydria, 91
acid-base evaluation, 188
acid, defined, 172
acids, metabolic sources, 180
acid phosphatase, 160
acrolein, formula, 27
acrylaldehyde, formula, 27
ACTH, 206
Addison's disease, 207
adenine, formula, 77
adenosine diphosphate, 84
adenosine triphosphate, 82
ADP, 84, 129, 150
adrenalcortical hormones, 206
adrenocorticotropic hormones, 206
adrenal medulla, 207
alanine, formula, 36
albinism, 130
aldohexose, 6
aldolase assay, 162
aldoses, 2
aldotetrose, 3, 4
aldotriose, 3

alkaline phosphatase, 160
alkaline tide, 89
alkaptonuria, 133
amidases, 72
amino acids
  α-, 38
  activation, 86, 141
  dehydrogenase, 126
  dimethylol, 40
  essential, 54
  glucogenic, 136, 147
  ketogenic, 136, 147
  oxidase, 126
  oxidative deamination, 126
  pool, 125
  salts, 39
  transamination, 126
aminoethyl sulfonic acid, 95
aminopeptidase, 92
ammonia, 127
amylase assay, 151
amylase, pancreatic, 92
amylase, salivary, 87
amylopectin, 22
amylose, 21
anabolism, description, 98
anaerobic glycolysis, 106
androgens, 210
anterior pituitary, 200
anterior pituitary hormones, 201
antidiuretic hormone, 202
antithyroid drugs, 204
arabinose, formula, 5
argentaffin cell tumors, 212

Index

arginine, formula, 37
  metabolism, 127, 128
ascorbic acid, 216
aspartic acid, formula, 36
  metabolism, 149
asymmetric carbon, carbohydrates, 3
atomic absorption flame photometer, 253
atomic absorption flame spectrophotometry, 253
ATP, 82, 129, 150
average, arithmetic, 260

base pairs, 138, 144
bases, defined, 172
  purine and pyrimidine, 80
  sources of, 181
Beer–Lambert Law, 224
Beer Law, 224
bicarbonate and carbonic acid measurements, 186
bile, 93
bile pigments, 93
bilirubin, 93
biliverdin, 93
bimodal distribution, 265
biotin, 220
blanks, photometry, 235
buffers, 175
buffer concentrations, 179
buffer dissociation constants, 177
buffer systems, biologic, 183
butyryl CoA, 118

calcium assay, 194
calibration curves, photometry, 239
carbohydrate
  alkaline metal reduction, 8
  description, 1
  fermentation, 7
  nomenclature, 4
carbon dioxide transport, 187
carboxypeptidase, 92
carcinoid, 212
carotenes, 214
catabolism, description, 98
catalyst, description, 56
catecholamines, 133, 207
cerebroside, 32
chloride measurement, 194
cholesterol, formula, 33
  metabolism, 95, 96, 121, 214
cholic acid, 95
choline, formula, 31
chylomicrons, 124
chymotrypsin, 91
chymotrypsinogen, 92

citrulline metabolism, 127
CoA, 116
CoASH, 115
coefficient of variation, 271
coenzyme A, 115
coenzymes, 59
colorimeters, 245
common ion effect, 174
competitive enzyme inhibition, 74
control sample, 272
corticosteroids, 206
cortisol, 206
CPK assay, 154
creatine kinase assay, 154
creatine metabolism, 128
creatine phosphate, 129
creatine phosphokinase, 154
creatinine, 129
cretinism, 204
Cushing's syndrome, 206
cutoff filter, 230
cuvets, 232
cyclopentanoperhydrophenanthrene, formula, 33
cysteine, formula, 36
cystine, formula, 36
cytosine, formula, 76
cytochromes, 149

dehydrogenases, 73
denaturation, description, 52
deoxyadenosine, formula, 78
deoxyguanosine-5'-phosphate, 78
5'-deoxyguanylic acid, 78
deoxynucleosides, names, 79
deoxynucleotidase, 92
deoxynucleotides, names, 79
deoxyribonucleic acid, 82, 137
2-deoxyribose, formula, 6
dextrorotation, 4
diffraction gratings, 231
digestion
  large intestine, 96
  mouth, 87
  small intestine, 91
  stomach, 88
dihydrosphingosine, 32
dihydroxyacetone, formula, 2
dihydroxyacetone phosphate metabolism, 103, 108
dipeptide, 44
1,3-diphosphoglyceric acid metabolism, 104
dissociation, constant, 170
dissociation, constants of buffers, 177
dissociation, of water, 172

Index

DNA, 82, 137
DNase, 92

electrolyte concentrations, serum, 192
electrolyte measurements, 193
electron transport system, 149
electrophoresis, 46
emission flame photometry, 249
enzyme
  action, 58
  active complex, 58
  activity measurements, 65
  activity units, 69
  coenzyme, 59
  cofactor, 59
  deficiencies, 113
  initial rate assay, 67
  kinetic assay, 67
  product, 58
  saturation, 65
  "steady state" assay, 67
  substrate, 58
  "two point" assay, 67
enzymes
  classification, 70
  history, 57
  hydrolases, 71
  inhibition competitive, 74
  inhibition noncompetitive, 74
  kinetics, 64
  "negative feedback," 109
  nomenclature, 69
  oxidoreductases, 73
  pH optima, 63
  product inhibition, 108
  proteolytic, 91
  reaction reversibility, 73
  specificity, 61
  temperature optima, 62
  transferases, 72
epinephrine, 133, 207
equilibrium, chemical, 165
equilibrium, constant, 167
erythrose, formula, 4
esterases, 72
estradiol, 209
estratriol, 209
estriol, 209
estrogens, 209
estrone, 209
ethanol amine, formula, 31

FAD, 116, 217
fat, depot, 115
  storage, 123
  transport, 123

fat-soluble vitamins, 213
fatty acids, essential, 30
  formulas and names, 25
  metabolism, 115, 147
  synthesis, 117
fatty liver, 123
filter, cutoff, 230
  interference, 230
  neutral density, 230
  transmission, 229
flame emission photometry, 249
flame photometer, 250
  "direct reading," 251
  "internal standard," 251
flame photometry, absorption, 253
flavin adenine dinucleotide, 116
flavoprotein, 149
flocculation, protein, 54
fluorescence, 247
fluorometer, 248
fluorometry, 247
FMN, 217
folic acid, 219
FP, 149
frequency distributions, 259
fructofuranose, formula, 17
fructofuranose-1-phosphate, formula, 18
fructosazone, 10
fructose, formula, 7
  metabolism, 103
fructose phenylhydrazone, 10
fructose-1-phosphate, formula, 18
fructose-6-phosphate, formula, 18
  metabolism, 102
furan, formula, 16
furfural, 11

galactopyranosyl glucopyranose, formula, 20
galactose, formula, 7
  metabolism, 101
galactosemia, 113
galactose-1-phosphate, 101
galactose-1-phosphate uridyl transferase, 101, 113
gallstones, 96
gastric acidity, 219
gastric analysis, 89
gastric juice, 88
genetic code, 140
glucaric acid, formula, 9
glucogenic amino acids, 136, 147
gluconeogenesis, 206, 207
gluconic acid, formula, 8
δ-gluconolactone, 112
δ-gluconolactone-6-phosphate, 111
glucopyranose, 15

280   Index

glucopyranosyl fructofuranoside, formula, 20
glucopyranosyl glucopyranose, formula, 19
glucosazone, 10
glucose
    α- and β-isomers, 14
    dehydrogenase, 112
    formula, 6
    metabolism, 99
    oxidase, 112
    phenylhydrazone, 9
glucose-1-phosphate, 99
glucose-6-phosphate, 99
    metabolism, 102, 108, 109, 113
glucosiduronic acid, formula, 9
glutamate oxalacetate transaminase assay, 153
glutamic acid, formula, 36
    metabolism, 149
glyceraldehyde-3-phosphate, 112
glyceraldehyde, formula, 1, 3
glycerin, formula, 25, 27
    metabolism, 108, 115
glycerin phosphate, 108, 119
glycerol, formula, 25
glycine, formula, 35
    metabolism, 95, 128
glycocholic acid, 95
glycogen metabolism, 99
glycogenesis, 113
glycogenolysis, 113, 206, 207
glycogen, storage, 113
glycolipids, 32, 122
glycolysis, description, 113
glycolysis, role of, 107
glycoside, 18
goiter, 204
GOT assay, 153
gout, 145
GPUT, 101, 113
growth hormone, 201
guanidinoacetic acid, 128
guanine, formula, 77

HCl, gastric, 88
HCl, titration, 89
heat shield, 232
helical configuration, 50
heme, 93
hemiacetal, 13
hexose monophosphate shunt, 109
5-HIAA, 135, 212
histidine, formula, 38
homocysteine, 128
homogentisic acid, 132
hydrochloric acid, 88
hydrocortisone, 206

hydrogen bond, 50
hydrogen ion concentration, 171
hydrolases, 71
β-hydroxybutyrate, 121
25-hydroxycholecalciferal, 34, 214
17-hydroxycorticosteroids, 207
5-hydroxyindoleacetic acid, 135
5-hydroxymethylfurfural, 12
hypergalactosemia, 114
hyperlipidproteinemias, 124
hyperthyroidism, 204
hypothyroidism, 203

imino acid, 37
"inborn errors of metabolism," 113
incident light, 223
interference filter, 230
intrinsic factor, 219
iodine number, 29
ionic bonds, 51
isoelectric pH, 46
isoelectric point, 46
isohydric mechanism, 181
isoleucine, 135

ketogenesis, 121
ketogenic amino acids, 136, 147
ketone bodies, 122
ketonemia, 122
ketonuria, 122
ketoses, 2
17-ketosteroids, 211
ketotriose, 3

lactase, 72, 92
lactate dehydrogenase assay, 155
lactate dehydrogenase isoenzymes, 156
lactate dehydrogenase isozymes, 156
lactic acid, 105, 146
lactose, 20
Lambert Law, 224
LD assay, 155
lecithin, formula, 31
leucine, 135
levorotation, 4
lipase assay, 152
lipase, pancreatic, 92
lipemia, post absorption, 123
lipids
    blood, 124
    description, 24
    iodine number, 29
    neutral, 24
    saponification number, 29
lipid transport, 123
lipoic acid, 220

lipoprotein, 32, 124
liver glycogen, 113
lysine, formula, 37
lyxose, formula, 5

magnesium assay, 197
malonic acid, in fatty acid synthesis, 117
malonyl CoA, 118
maltase, 72, 92
maltose, formula, 19
maple syrup disease, 135
mean, 260
median, 261
meiosis, 138
melanins, 130
menadione, 216
messenger RNA, 139
metabolic controls, 108
metabolism, definition, 98
metanephrine, 133, 207
methionine, 128
3-methoxy-4-hydroxymandelic acid, 133, 207
milk sugar, 23
mitosis, 137
mode, 261
molar absorptivity, 240
monosaccharides, 19
mRNA, 139
multimodal distribution, 265
muscle glycogen, 113

NAD, 59, 149, 218
NAD, formula, 86
NAD, reduction mechanism, 86
NADP, 86, 149, 218
negatively skewed distributions, 264
nepelometer, 247
nepelometry, 246
neuroblastoma, 133, 208
neutral density filter, 230
neutral lipids, 24
niacin, 218
nicotinamide, 218
nicotinamide adenine dinucleotide, 59, 149
nicotinamide adenine dinucleotide, formula, 86
nicotinamide adenine dinucleotide phosphate, 86, 149
ninhydrin, formula, 41
  reaction with amino acids, 41
noncompetitive enzyme inhibition, 74
norepinephrine, 133, 207
normal distribution, 262
normetanephrine, 133, 207
nucleases, 72
nucleic acid synthesis, 144

nucleosides, 77, 79
nucleoside polyphosphates, names, 84
nucleotides, 78, 79
number rounding, 258

oleic acid, formula, 26
oligo-1,6-glucosidase, 92
optical isomers, carbohydrate, 3
ornithine metabolism, 127
osazone, 10
osteomalacia, 214
oxidative phosphorylation, 150
oxalacetic acid, 149
oxidoreductases, 73
oxygen transport, 187
oxytocin, 202

pancreatic juice, 91
pancreatic secretion, 91
pantothenic acid, 218
papain, 62
parathormone, 211
parathyroid glands, 211
parathyroid hormone, 211
partial pressure, gases, 187
PBI, 205
P_{CO_2} determination, direct method, 190
P_{CO_2} determination, equilibrium method, 189
pellagra, 218
% T, 225
percent transmittance, 225
pentoses, 109
pentosuria, 114
pepsin, 89
pepsinogen, 89
peptide bond, 42
peptide link, 42
pH
  defined, 171
  regulation, lungs, 183
  regulation, kidney, 184
  measurements, 187
phenylacetic acid, 132
phenyllactic acid, 132
phenylalanine, formula, 37
  metabolism, 130
phenylalanine, hydroxylase, 130
phenylketonuria, 132
phenylpyruvic acid, 132
pheochromocytoma, 133, 208
phosphatase assays, 158
phosphate assay, 197
phosphatidic acid, formula, 30
  metabolism, 119
phosphatidyl choline, formula, 31
phosphatidyl ethanolamine, 120

phosphocreatine, 129
6-phosphofructose, formula, 19
phosphoglucomutase, 101
6-phosphogluconic acid, 111
6-phosphogluconolactone, 108
3-phosphoglyceraldehyde, metabolism, 103
2-phosphoglyceric acid, 105
3-phosphoglyceric acid, 104
3-phosphoglycerin, 108
phospholipids, metabolism, 119
phosphomonoesterase, 158
phosphorolysis, 100
photodetector, 233
photometry
  calibration curves, 239
  emission flame, 249
  flame absorption, 253
  quantitation, 237
photomultiplier, 234
photoresistive photodetector, 234
phototube, 234
photovoltaic photodetector, 234
polydeoxynucleotide structure, 82
polynucleotide structure, 81
polyribosomes, 141
polysaccharides, 21
polysomes, 141
positively skewed distribution, 264
posterior pituitary hormones, 202
potassium assay, 193
precision, 269
pregnanediol, 210
prisms, 230
progesterone, 209
proline, formula, 37
proteases, 72
protein
  acidic, 45
  basic, 45
  C terminal, 45
  N terminal, 45
  structure, primary, 48
  structure, quaternary, 51
  structure, secondary, 50
  structure, tertiary, 51
protein-bound iodine, 205
proteolipid, 32
proteolytic enzymes, 91
purine, derivatives, names, 79
  metabolism, 145
  formulas, 77
pyran, formula, 15
pyridoxal, 218
pyridoxal phosphate, 218
pyridoxamine, 218
pyridoxine, 218

pyrimidine, derivatives, names, 79
  metabolism, 145
  formulas, 76
pyrophosphate, 85
pyruvic acid, 105, 149

radiant energy, sources, 228
  spectrum, 222
random coil, 50
reaction velocities, 166
readout device, photometry, 235
reproducibility, 270
riboflavin, 116, 217
ribonuclease, 92
ribonucleic acid, 82
ribonucleosides, 110
ribonucleotides, 110
ribose, formula, 5
ribose-5-phosphate, 110, 112
ribitol, 217
ribulose-5-phosphate, 110, 114
rickets, 214
RNA, 82
RNase, 92

saliva, function, 87
saponification, 27
  number, 29
serine, formula, 31, 36
serotonin, 135, 212
serum glutamate oxalacetate transaminase assay, 153
SGOT assay, 153
significant figures, 257
Simmond's disease, 207
skewed distributions, 264
soaps, 27
sodium assay, 193
Sörensen formal titration, 40
specific absorptivity, 240
spectrophotometers, 245
sphingomyelins, 31
sphingosine, formula, 31
standard deviation, 265
standardization curves, photometry, 239
standard solution, 272
starches, 22
sterioisomerism, carbohydrate, 3
steroid, 33
sterols, 33
succinic acid dehydrogenase, 75
sucrase, 72, 92
sucrose, formula, 20

T, 225
T_3, 203

Index   283

T_4, 202
taurine, 95
taurocholic acid, 95
TBG, 205
testosterone, 210
threose, formula, 4
thiamine, 217
thiamine pyrophosphate, 217
2-thio-6-propyluracil, 204
thiouracil, 204
thiourea, 204
thymine, formula, 76
thyroid, 202
    thyroid hormone assays, 205
thyronine, 203
thyroxine, 134, 202
thyroxine binding globulin, 205
o-tolidine, 112
o-toluidine, 11
transaminases, 72, 126
transamidination, 128
transcription, in protein synthesis, 139
transfer RNA, 141
transferases, 72
translation, in protein synthesis, 143
transmethylation, 72, 128
transmission filter, 229
transmittance, 225
transmitted light, 223
transphosphorylation, 72
tricarboxylic acid cycle, 147
triglyceride, description, 26
    metabolism, 115, 119
triiodothyronine, 203
trioelein, formula, 28
tripalmitin, formula, 25
tripeptide, 44
tristearin, formula, 27, 28
tRNA, 141
trypsin, 91
trypsinogen, 91
tryptophan, formula, 38
    metabolism, 135
TSH, 203

turbidimetry, 246
turbidometer, 247
tyrosinase, 130
tyrosine, formula, 37
    metabolism, 130, 134

UDP-galactose, 101
UDP-glucose, 99, 101
uracil, formula, 76
urease, 61, 72
urea synthesis, 127
uric acid, 145
uridine diphosphogalactose, 101
uridine diphosphoglucose, 85, 99, 101
uridine, formula, 77
uridine triphosphate, 85
urobilinogen, 93
UTP, 85

vacuum phototubes, 234
valine, 135
    formula, 36
vanilylmandelic acid, 133
vitamin
    A, 213
    B_{12}, 219
    C, 216
    D, 214
    D_3, 215
    D, formula, 34
    definition, 213
    E, 214
    K, 216
    supplements, 220
VMA, 133

water-soluble vitamins, 216
waxes, 30

xylose, formula, 5
    metabolism, 114

zwitterion, 39